Invisible Mind

Invisible Mind

Flexible Social Cognition and Dehumanization

Lasana T. Harris

The MIT Press
Cambridge, Massachusetts
London, England

This book was set in Stone Sans and Stone Serif by Toppan Best-set Premedia Limited. Printed and bound in the United States of America.

Library of Congress Cataloging-in-Publication Data

Names: Harris, Lasana Tacuma, 1981- author.
Title: Invisible mind : flexible social cognition and dehumanization / Lasana T. Harris.
Description: Cambridge, MA : MIT Press, [2017] | Includes bibliographical references and index.
Identifiers: LCCN 2016032131 | ISBN 9780262035965 (hardcover : alk. paper)
Subjects: LCSH: Social perception. | Emotions and cognition. | Humanity--Psychological aspects. | Evolutionary psychology. | Brain--Evolution.
Classification: LCC BF323.S63 H37 2017 | DDC 302/.12--dc23 LC record available at https://lccn.loc.gov/2016032131

10 9 8 7 6 5 4 3 2 1

To Lisa-Marie and Lyon

Contents

Prologue

A young man sits on a bench at a street corner. Another young man walks by, stops, turns toward the first and pulls a gun. Suddenly, two more young men come running, one from each direction, guns blazing. The victim is now in full flight, attacking one of his assailants in a struggle for his life. But alas, a few seconds later, he lies dead in a ditch at the side of the street, one of his assailants standing over him plugging bullets into his motionless body to ensure his fate.

A toddler wanders onto the balcony of her sixth story apartment. Her parents, occupied in the living room, do not notice her attempt to squeeze through the narrow gap in the outer balcony wall. After a few attempts, she makes it, and now dangles from the ledge. A stranger below sees the toddler, and quickly begins climbing the ledges of the apartments below. He grabs hold of the toddler just before she falls, hands her over to her now frantic parents, and walks away to a chorus of "Thank you."

Imagine that the same person was involved in both of the above scenarios as assailant and hero. How is it possible for someone to be both tender and violent? How can a person kill people during the day yet be kind and caring toward her own family at night? This book attempts to explain everyday events that appear on the front pages of newspapers and websites, events that depict horrid human behavior, including terrorist attacks, racially motivated shootings, human trafficking, violent intergroup conflicts, and historical incidents of atrocities such as New World slavery and genocide. It does so by taking a social neuroscience approach: by integrating theories from a variety of disciplines to explain *how human beings flexibly engage social cognition and, thereby, flexibly perceive humanity*. Stated differently, this book examines *dehumanized perception*—the withholding of social cognition (thinking about someone else's mind) to some people (see Harris & Fiske, 2009; Lee & Harris, 2013, for review). Specifically, it explores *how* we engage social cognition, and when we do not, enabling social interaction and subsequent intra- and intergroup contexts. Furthermore, it addresses *why* the brain might engage or withhold social cognition. I argue that flexible social cognition allows for a human/nonhuman distinction. To explain this phenomenon, I integrate theory in social, cognitive, and developmental psychology, philosophy, evolutionary anthropology, and economics with neuroscience data to describe this typically human process and explore its brain correlates. Each chapter describes research in different disciplines, integrating established theories within each discipline in an attempt to provide a parsimonious perspective on this fundamental human process. I also describe biological structure and processes that underlie this phenomenon, including the brain, as well as possible genetic, hormonal, and physiological mechanisms.

At this point a critic may wonder whether the central argument is tautological or circular: does the human/nonhuman distinction equate to the greater or lesser engagement of social cognition? If so, does this imply that the central thesis can be reduced to "the degree of engagement of social cognitive processing results from the flexibility of social cognitive processing?" I argue that social cognition engagement is used to determine whether a target is human or not. For instance, I might determine that an animal that flies must have wings. However, wings are not reserved just for animals that fly. Similarly, a target that is human must engage social cognition, but social cognition engagement is not just reserved for humans. The central argument is that people have the ability to regulate whether they think about the minds of humans (engage social cognition). This ability is evolutionarily preserved perhaps because it proved functionally useful. Specifically, human beings evolved in much smaller groups than typical modern cities, states, and countries. Thus, migration between these smaller groups required flexible social cognition because people are capable of deception. Furthermore, the social context provides cues as to whether social cognition would be useful or not. However, this ability may prove maladaptive in modern societies where group size has expanded. Nonetheless, people use this engagement as an index of whether a target is considered as human: more engagement makes a target more likely to be human, whereas less engagement restricts the possible humanity of a target.

The sentiment I discuss is similar to existing theories in evolutionary anthropology: primarily the social intelligence (Humphrey, 1976) and the social brain (Stiller & Dunbar, 2007 hypotheses. The former argues that the human brain evolved because of the hypersocial nature of human beings. Such social networks and the resulting problems drove human intelligence to an advanced stage beyond other species. Evidence for this hypothesis is plentiful, but the most compelling demonstrates that our closest genetic ancestors (chimpanzees and bonobos) can solve some physical tasks as well as humans but struggle with most complex social tasks. Of course I am not saying that other great apes can build cars, computers, and civilizations as complex as humans, but they can figure out how to use tools to reach food stuck inside of tubes, for instance. In one such study, five- and six-year-old children, and chimpanzees, were shown how to use a stick tool to access food in a tube (Horner & Whiten, 2007). The experimenter modeled successful acquisition of the food using the stick tool but often included unnecessary actions or errors with the tool that did not result in the food reward. While the children often imitated these errors, at least some of the chimps successfully obtained the food reward without committing the errors. Such evidence suggests that chimps are efficient tool users, and this efficiency extends to other tools besides sticks, such as using stones to crack nuts (see Seed & Byrne, 2010 for a review).

The social brain hypothesis argues that human brains can only process 150 people within one meaningful social network; perhaps this was the maximal size of social groups that facilitated the brain's evolution. Evidence for this comes from considering the social complexity hypothesis, which argues that the complexity of information (defined as the number of premises in a logical argument) that people can process limits the size of social groups that an organism can successfully maintain. Computational models have provided initial support for these hypotheses: larger group size can only be achieved by processing more complex information; this is not the case for smaller group size (Dávid-Barrett & Dunbar, 2013).

The latter ideas regarding social group size limits fit nicely with social psychological theory describing the cognitive burdens of social information processing (see Fiske & Taylor, 1984, 1991, 2013 for an early account). For instance, each person has a number of identities and personality traits that serve as premises for plausible behavior the person might enact. If George is shy, nurturing, smart, reserved, brooding, caring, and fun-loving, and I had to represent each of these personality traits each time I encountered George because any of these traits might predict his actual behavior, then I might be cognitively overloaded (especially if I did this not only for George but for all other people I met with in a social context). Keeping in mind this vast amount of information for each person encountered would lead to extreme cognitive load. The cognitive miser hypothesis (Fiske & Taylor, 1984, 1991, 2013) argues that *social heuristics* such as *categorization* and *stereotyping* occur to save cognitive resources. More recent accounts support these ideas, both with behavioral (Waytz, Epley, & Cacioppo, 2010) and with brain data (Lee & Harris, 2014b). Furthermore, and most importantly, social psychological research underscores the role of social context (including other people) in shaping behavior since the inception of the discipline (Lewin, 1935). This is consistent with both the social intelligence and social brain hypotheses because other people play an important role both in presenting challenges that require more sophisticated intelligence to overcome and in burdening brain networks, potentially maximizing the contribution of billions of neuronal connections.

In short, perceiving and interacting with other people require substantial brain processing, something humans engage in spontaneously without effort or intention. The resulting high cognitive load remarkably differs from other types of high cognitive load, such as the demands placed on working memory by nonsocial cognition, which famously maxes out at seven plus or minus two numbers (Miller, 1956), as well as other abstract cognitive loads related to tasks, but including visual working memory and attention (see Botvinick & Cohen, 2014). Those forms of cognitive load seem to be much more burdensome than social information processing. This consideration not only brings into question dual-process accounts of cognition but isolates cognitive load (or rather cognitive activity) from social-related brain activity, casting the latter as somehow different or special. This special status attributed to social cognition is also common in both evolutionary anthropology hypotheses.

So what is the added theoretical contribution of this book beside the fact that it comes from a social neuroscientist? What differentiates the account of social cognition provided herein from either the social intelligence and/or social brain hypotheses? The social intelligence hypothesis argues that the brain's evolution was driven by difficult social problems. Given that human social groups are much larger than those of other species, the human brain evolved to be more complex to solve a wider variety of social problems. I agree with the impact of the social context but not with the view that increasing group size and the resulting increasingly complex coordination problems have been the driving forces behind the brain's evolution. Instead, I argue that deception (which can be present in both smaller and larger social groups) coupled with group migration have presented a complex problem, requiring flexible social cognition. I am agnostic as to whether this is a form of intelligence but view it as functionally necessary to survive in an unpredictable and changing social world. Here, the theories differ on the source of complexity that potentially drives the social brain's evolution.

The social brain hypothesis describes the brain as attempting to solve complex coordination problems. It focuses on the social group and describes the brain as constrained. However, I focus on the single individual, not the social group, and describe the impact that solving complex problems involving each individual might have had on the brain, *flexible social cognition*.

Therefore, I do not view the brain as limited because it does not consider all social connections simultaneously (indeed, people may not consider their entire social network), but I acknowledge that flexible social cognition does not allow for behavior with much larger social groups than those present during the human brain's initial evolution. Additionally, heuristic methods, an evolved feature of the system to enable processing as the numbers exceeded capacity, are fundamentally flawed since they rely on Bayesian[1] processes. Thus, flexible social cognition can be destructive or transformational, allowing the continued flourishing of humanity with obvious costs.

This idea of flexible social cognition hinges on a thought experiment I describe in the sixth chapter. Essentially, it asks how one could evaluate someone's intentions in the absence of useful information beyond that which the person provides? If a person could be beneficial and/or harmful when approached, but approach motives (hinged on survival) were equivalent to avoidance motives (also hinged on survival), how do you decide which action to take? What information would you ask people to provide when they are the only source of information given that a handy tool social intelligence has evolved is deception? The exploration of this question results in a focus on the social context, and it demonstrates the importance of this factor in actually driving people's behavior (facilitated by the social brain).

A Social Neuroscience Approach

Before we go further, it is important to define the *social neuroscience approach* used throughout. Although I provide a brief guide to the brain in the appendix, I have included this section not simply for the benefit of the non-neuroscientist reader but to clearly identify my approach to the social behavior questions presented above. Social neuroscience holds the promise of addressing complex issues such as racially motivated police violence, genocide, terrorism, financial collapses, poverty, structural racism and discrimination, intergroup conflicts, healthcare and educational disparities, and other social ills that plague society. The neuroscience revolution has touched a vast array of academic disciplines; from philosophy to physics, it seems that many academic disciplines become rejuvenated when considering the brain. Neuroscience allows a glimpse into the proverbial black box of cognition and serves to generate novel theories about how people do what they do. This approach allows the scientist to avoid reductionism while it permits scientists to dig deeper into the biological mechanisms driving human behavior. The intention of the book is not to go into depth on any single topic but, rather, to cover a vast array of topics simultaneously in the hope of answering a single, focused question related to human behavior. Stated a second time, the aim of the book is not to review a vast literature or to delve deeply into a very narrow topic but, rather, to take a broad interdisciplinary approach, *a social neuroscience approach* covering many topics, in the hope of making progress on a single focused question related to flexible social cognition.

However, such an approach relies on a clear, indisputable understanding of brain function. At the moment, no such understanding exists. Much of this book, then, will describe a person perception and/or social cognition brain network (for review, see Amodio & Frith, 2006; Frith & Frith, 2001; Gallagher & Frith, 2003; Haxby, Gobbinni, & Montgomery, 2004; van Overwalle, 2009). This includes areas of the brain that are more active when people perceive or think about other people rather than objects, such as parts of medial prefrontal cortex (MPFC), superior temporal sulcus (STS), temporal-parietal junction (TPJ), precuneus, posterior cingulate (PCC), the amygdala, and the insula. Although none of these regions independently can be considered "social" because each responds to nonsocial stimuli as well, together their activation suggests the brain has engaged a social task or function; they provide an index of the cognitive process. However, inferring psychological phenomena from patterns of brain activity relies on reverse inferences, which present a challenge for epistemology. Specifically, each brain region underlies multiple functions; therefore, brain activation can reflect any of a number of psychological processes, making identifying a sole process nearly impossible. As a result, I rely on empirical evidence in disciplines related to neuroscience to provide converging evidence.

The concept of social neuroscience used herein diverges from other social neuroscience approaches popular in academia. Considering social neuroscience as a merger between social psychology and cognitive neuroscience presents a number of challenges. These challenges can best be thought of as cultural differences between academic disciplines (see Lee & Harris, 2014a for a similar discussion). For instance, social psychologists prefer between-subjects paradigms (study designs in which a participant is exposed to only one experimental condition) because they alleviate demand effects by concealing the experimental manipulation. Moreover, the complexity of many social psychological paradigms demands a between-subjects design because it is not psychologically possible for participants to be in the different conditions of the design. These designs allow social psychologists to study phenomena beyond the range of cognitive neuroscience because they can be applied to a broader range of social phenomena and behavior, including topics rarely studied in traditional cognitive neuroscience such as close relationships. Absent from a view of such behavior are data regarding minute biological changes across the person, and changes in cognitive state must be inferred.

However, cognitive neuroscience prefers within-subjects designs (study designs where each participant is exposed to all experimental conditions) because of the signal-to-noise issue inherent in physiological measurement. Physiological measurements are noisy; that is, any signal reflecting behavior is buried within noise produced from the actual physiological measure. For instance, facial electromyography (EMG), a physiological method that measures the minute movement of facial muscles underlying facial displays of emotion, requires placing tiny electrodes on the face. These electrodes require a conducting gel between the electrode surface and the skin. Electrical signals emanating from the muscles must pass through all the layers of skin, hair on the face, dead skin cells, and the conductive gel before getting to the surface of the electrode. This journey diminishes the signal. Moreover, these electrodes also detect other electrical currents in the environment, including those emanating from computer screens, some common forms of lighting, and electrical outlets. Therefore, the electrical signal from the muscle is buried within a plethora of other electrical signals. Thus, researchers rely on averaging across a large number of trials in order to better detect the electrical signal, improving the signal-to-noise measure. Such a strategy requires increased power (the ability to detect an effect), which is gained by employing a within-subjects design. This signal-to-noise problem is common to all physiological measures.

As a result, cognitive neuroscientists focus on a very different type of data and aspect of human behavior. These scientists are able to detect biological change concurrent with human behavior, although the range of human behavior is impoverished. Hence, the choice of experimental paradigm using this approach to social neuroscience is fraught with difficulty because cognitive neuroscientists and social psychologists are often studying past each other at different levels of human behavior, and any compromise does not fully address the other discipline's concern borne out of cultures where scientists peer worriedly over their computers obsessing primarily about maximizing signal and eliciting ecologically valid behavior, respectively.

And the problems simply worsen. Social psychologists' broad view of human behavior has led to a careful description of complicated psychological concepts, studying specifically dispositional attribution, impression formation, person perception, stereotyping, thin-slicing, spontaneous trait inferences, the fundamental attribution error, correspondence bias, actor-observer effect, and self-perception all as social cognitive processes. But in the brain, research has not yet localized each specific process to a specific network of brain regions or pattern of activity. As such, cognitive neuroscientists study a "social cognition network" where all of these related, yet nuanced, processes occur. The fine resolution necessary to dissociate such nuanced phenomena is not yet available for examination in the awake, normally functioning human brain. Therefore, cognitive neuroscience sticks to the broader psychological concepts, conflating phenomena. These problems do not render this form of social neuroscience invalid, but they do limit the conclusions one can draw from evidence collected with this methodology if one only considers such evidence in isolation. Stated differently, the broad focus of social psychology results in minute psychological processes, whereas the signal-to-noise focus of cognitive neuroscience results in broad categorization of concepts. Therefore, social psychologists and cognitive neuroscientists are considering different aspects of human behavior because of their different sources of data and experimental manipulations.

Perhaps as a consequence, the most popular approach to social neuroscience views it as simply a cognitive neuroscience study of social behavior. This approach classes social cognition with other cognitive phenomena, including memory, attention, emotion, decision making, and cognitive control. Therefore, social cognition with its resultant behavior is simply a form of cognitive processing where social stimuli are used instead of nonsocial stimuli. Fundamentally, this approach aims to better understand the brain and cognition as topics of inquiry. With this goal, social neuroscience is a useful approach to understanding the brain because the brain evolved in a social context, and social phenomena best instantiate brain processing. Here, I abstain from this approach because my desire is to understand social behavior, not just the brain and how it instantiates social cognition. Given my abstention, neuroscience becomes but another tool in a repository that includes the other disciplines beyond cognitive neuroscience that I discuss throughout this book, including philosophy, economics, legal and policy studies, developmental psychology, evolutionary anthropology, and, most importantly, social psychology. Social psychology holds a status of primary importance because I use social psychological theory as a foundation for understanding social behavior. This does not mean that I am not interested in brain function, but my interest in brain function persists only to the extent that it sheds light on the mechanisms of social behavior.

An alternate approach focuses on social behavior, using the brain as a point of convergence for psychological processes. In this alternate approach the brain is an additional dependent variable or index of social processing, and insights about social behavior can be gained by knowing how it engages the brain. However, despite the term *social neuroscience,* this alternate approach is not limited simply to the brain, and insights can be gleaned by looking at body physiology outside the central nervous system provided by hormones and genetics, across species, and in related academic disciplines such as developmental psychology and behavioral economics. Although the prevailing approach may also integrate these bits of empirical data, most importantly this alternate approach focuses on the social processes. This makes the possibility of solutions to societal problems more likely since they occur at the psychological and systemic levels, not the level of the brain. This alternate approach to social neuroscience is the one employed in this book.

Caveats

Before proceeding, it behooves me to note that I am not an expert in all of the disciplines that I discuss in this book. I have been trained as a social psychologist and cognitive neuroscientist and consider myself a closet philosopher at best, so my expertise in the other disciplines mentioned comes almost completely from long talks and general guidance from my wonderful collaborators and colleagues Rachel Kranton, Jingzhi Tan, Walter Sinnott-Armstrong, and Ophelia Deroy, themselves experts in economics, evolutionary anthropology, and philosophy; each one was more than willing to sit with a social neuroscientist and help him to understand basic concepts in her or his own discipline relevant to social cognition. Even in my own fields I rely largely on the theory of my graduate and postdoctoral mentors Susan Fiske and Elizabeth Phelps in social psychology and cognitive neuroscience, respectively, each armed with sufficient theory based on decades of research that is applicable to flexible social cognition. Therefore, rather than consider this book a wise summation of a career of research, or a review of a vast literature, I ask that the reader consider it a unique social neuroscience attempt—the applying of neuroscience data to interdisciplinary research questions guided by a philosophical approach. By engaging in this exercise, I simply hope to stimulate discussion among scholars and students across the variety of disciplines and to clarify my own thinking on complex social problems. Finally, three anonymous reviewers helped shaped the final version of this book; I owe them my gratitude as well.

My second caveat surrounds integrating so many disciplines. I try to speak across disciplines to scholars and students who use different jargon and reside within different disciplinary cultures. To do so, I must speak plainly, but at times I may use psychological jargon unfamiliar to the reader. In these cases I refer the reader to the glossary or to any introduction to psychology textbook. Additionally, to tell a story across these disciplines, I may at times make an inferential or logical leap. Finally, the discussion may at times seem spastic or tangential as I try to integrate knowledge across a vast array of disciplines and topics. When any of these problems occur, I ask humble forgiveness from the reader. I only hope these shortcomings stimulate debate and scientific exploration.

Ten Chapters

The first chapter provides a philosophical platform for the book, elucidating the research question and epistemology. It updates an equation at the root of social psychological theory that emphasizes the interaction of the person and the social context in driving behavior. Next, this chapter defines the terms relevant for navigating the theory proposed within the book, specifically *mental state inferences* as a form of social cognition, *flexibility*, and *social groups*. This chapter dissociates the specific use of these terms from related concepts and uses arguments in philosophy of mind, philosophy of science, and epistemology to explain why these specific, narrow definitions are appropriate and useful.

The second chapter explores examples of social cognitive ability in nonhuman primates and other animals, comparing and contrasting these abilities with those in humans. This provides an explanation of why flexible social cognition may have occurred and been preserved throughout our evolutionary history. It also describes how social cognition employed in humans and in other animals shapes interactions and allows agents to influence each other. The advantage gained looking across species comes from knowing what traits a common ancestor held. This homology answers the *how* question of social cognition: "How did humans evolve the trait?"

Looking across species also tell us what traits developed since humans' divergence from their closest cousins due to the uniquely human environment. This analogy answers the *why* question: "Why did we develop the trait?" This chapter explores both homology and analogy, looking at shared social cognition abilities with our closest ancestors, nonhuman primates such as chimpanzees and bonobos, as well as shared social cognition abilities with species not as closely related but who live in human environments such as domesticated animals like dogs. In so doing, the second chapter speculatively addresses possible ultimate and proximal causes that drove the evolution of social cognitive abilities, examining how survival and reproduction required social cognition in humans and examining other psychological mechanisms including specific emotions such as fear that may have allowed social cognition to evolve.

The third chapter addresses prediction as a motive for social cognition. This allows for a discussion of one of the tenets of social cognition, namely that it is useful for predicting the behavior of other people. Since people's behavior is driven by motives, goals, desires, intentions, and other hidden mental states that cannot be seen, only inferred, social cognition is useful for predicting the behavior of others. This chapter explores the role of prediction in survival, describing prediction from a Bayesian perspective. It also presents behavioral economic theory, and explores neuroeconomic evidence for the brain correlates of social prediction and decision making. Following a discussion of the proximate psychological mechanisms, the chapter also explores social prediction violation before a discussion of social reward and punishment. I explain these phenomena and argue that integration of Bayesian approaches is required to describe how learning and prediction may have necessitated social cognition to be flexible in order for humans to thrive in complex social environments. This chapter takes an interdisciplinary perspective that encompasses emotion, behavioral economics, and learning to describe how the need to predict the behavior of others may have spurred the evolution of social cognition. Most interestingly, it concludes by questioning whether social cognition is necessary for predicting others' behavior.

The fourth chapter addresses the second motive for social cognition—explanation. Since people are interested in satisfying epistemological needs such as understanding why things occur and why people may do the things they do, social cognition is a handy tool for explanation. This chapter describes the need for understanding as a fundamental human motive.[2] It discusses classic developmental and social psychological evidence for anthropomorphism and imagination in children and adults, as well as dehumanized perception, as examples of social cognitive processes that satisfy the need for understanding, and it describes the utility of such phenomena for survival. Understanding is a core social motive presumably linked to the evolution of consciousness. Human beings are motivated to understand why the things around them occur, and social cognition provides a ready explanatory mechanism. For instance, because mental state inferences are inferences, they utilize the tools of formal logic, allowing people to draw rational conclusions from existing priors. Mental state inferences thereby facilitate explanation not simply of human behavior but of all meaningful action. This chapter reviews cases of anthropomorphism and dehumanization, comparing effusive magical thinking[3] (anthropomorphism) to reduced magical thinking (dehumanization), and it explores a paradox: the tendency to think magically reduces as inferential and logical reasoning abilities strengthen. Incidentally, anthropomorphism also demonstrates flexible social cognition because it captures social cognition stretched to entities that do not have mental states. Here I discuss how cultural norms define the space where and to whom magical thinking specifically, and social cognition more generally, are acceptable.

The fifth chapter begins the discussion of social cognition engagement as useful for a human–nonhuman distinction by presenting social and developmental psychological evidence that adults and children integrate information using Bayesian rules to infer people's mental states. This chapter thus begins a discussion that relates flexible social cognition to a human–nonhuman distinction. It also describes social cognition as an affective-cognitive psychological process and presents social neuroscience evidence about the brain correlates of inferring something about someone else's mental state. Given the dual functions of prediction and explanation, social cognition provides necessary information to promote survival in social contexts. The inference process itself relies on probabilistic estimations based on base rates, priors, and expected outcomes. As a result, this process is highly logical, often likened to scientific hypothesis testing. This chapter reviews the developmental psychology literature supporting this naive science strategy that facilitates social cognition. Specifically, it focuses primarily on work with infants and children documenting their human–nonhuman distinctions. During this discussion, I consider why some social cognitive abilities may be innate (or at least come online very early in life), whereas others require development of complementary psychological abilities. Thus, the fifth chapter summarizes and integrates the research described in the previous three chapters, setting the stage for the next two chapters, which posit the main theoretical contribution of this book.

The sixth chapter posits that flexible social cognition may have been necessary to facilitate the evolution of a highly social species. This crucial chapter argues that, given its perseverance throughout human evolution, flexible social cognition and the resultant human–nonhuman distinction may have driven our species forward toward modernity. This chapter integrates theory in social psychology and behavioral economics to describe how flexible social cognition enabled modern humans and presents explanations that may account for both pro- and antisocial behavior. The discussion in this chapter is framed by a thought experiment modeled on a possible scenario in our evolutionary history. Thus, the chapter provides an introduction for all of the subsequent discussions that will guide the outcome of the thought experiment. During the discussion of the hypothetical scenario, I utilize the concepts of flexibility and mental states to posit a unique theory: specifically, flexible social cognition facilitates distinctions between intentional and nonintentional agents that extended to differentiate helpers and harmers (or hinderers as some researchers describe them). I then differentiate this theory from related theories that attempt to explain why human beings create social groups and how social cognition contributes to social group formation.

The seventh chapter describes the phenomenon of reduced spontaneous social cognition engagement in the presence of other people—dehumanized perception—within the framework of a larger discussion about the power of the social context. This chapter uses dehumanization as one example when social cognition, through its flexibility, is withheld, and it begins the discussion of consequences of flexible social cognition. The chapter discusses social neuroscience and social psychological evidence for dehumanized perception and describes how it may have been adaptive for different social contexts.

Building from the previous chapter, this chapter describes evidence that supports the unique theory. Moreover, this chapter describes the impact of social context on flexible social cognition, using empirical data to support claims that social contexts may shape group formation. In brief, this chapter argues that social contexts necessitate flexible social cognition, thereby allowing people to use this flexibility to adapt to a variety of social contexts, each with specific rules that reward different behaviors, thus recharacterizing the metrics of successful social interaction. Therefore, this chapter provides an overarching framework for the next two chapters, which describe specific instances of social context effects on flexible social cognition in detail.

The eighth chapter describes how economic trade benefits from flexible social cognition. This chapter serves as the first example detailing how flexible social cognition operates in a specific context. As a result, it revisits the discussion of prediction in chapter three and focuses on economic decisions. It presents social neuroscience evidence and examines how the evolution of barter and trade may have also promoted flexible social cognition. This chapter describes my own recent work and the work of others documenting a social contextual impact on flexible social cognition in economic markets. This chapter relates social cognition to economic exchange and describes the former as necessary to successfully complete the latter. In short, social cognition provides information about the intentions of both parties in the exchange, validating pricing and valuation processes. This chapter builds from classic economic theories and demonstrates how the implications of my unique theory are consistent with many of these predictions and how they can further these theories simply by consideration of additional social variables traditionally thought to be irrelevant in classical economic theories.

The ninth chapter describes how enforcing the social contract may also necessitate flexible social cognition. This chapter provides the second example of how flexible social cognition operates in a second social context. It presents social neuroscience evidence and examines how moral rules and social norms may have shaped flexible social cognition. As a result, the chapter revisits the discussion of norms that began in chapter three, but now applies them to the legal context. Thus, chapter nine covers morality, the law, and the social context. The three are inextricably linked since laws often reflect both moral rules (such as natural laws that prohibit killing, although such behavior is legally acceptable under particular circumstances) and social norms (such as contributions toward a public good often made via taxation) within a given culture. This chapter describes my own recent work and the work of others documenting bias in legal decision making due to the exclusion of some persons from the social contract. I focus on racial bias in the United States as an example and discuss how one instantiation of that bias—fatal police shootings of black males in the United States—may result from this exclusion. Therefore, this chapter explores the legal arena as a second social context that may impact flexible social cognition.

The tenth chapter speculates about further implications of the research described and theory posited in the book for how human beings may interact in healthcare, policy, and intergroup contexts such as athletic competitions and international disputes. This chapter serves to tie a bow around the discussion of flexible social cognition, highlighting questions that remain unanswered and possible areas of exploration for future research. In this regard the book's final chapter builds from the previous three chapters, probing how flexible social cognition may contribute to current real world problems and advances. It then considers what the next hundred years of humanity may face given recent technological advances that allow people to infer the mental states of others whom they have never met or interacted with in a face-to-face manner typical of all of human existence to this point in time. This most important chapter guesses at the consequences of these recent developments, offering suggestions for humanity moving forward that may allow us to all live in harmony, grounded in empirical research and theory.

Notes

1. Bayesian: Bayes' theorem was developed in the mid-eighteenth century as a way of using new evidence to update beliefs. It is often described as nothing more than common sense reduced to calculation. In short, Bayes' theorem is as follows:

p(a|b) = p(b|a)*p(a)/p(b), where

p(b|a) = likelihood,

p(a) = prior,

p(b) evidence.

p(a|b) is often described as the posterior, and p(b|a)*p(a) is described as the model. This formula is commonly used today in psychology, neuroscience, behavioral economics, and other disciplines attempting to explain behavior. It is often implied that people employ Bayesian calculation during person perception/social cognitive processes, and I make this assumption in this book as well.

2. Social psychological theory often describes core human motives: broad goals that drive behavior. One popular account describes BUCET motives: belonging, understanding, controlling, enhancing the self, and trust (Fiske, 2003). Belonging describes the human need to establish and maintain social connections, often accomplished via forming and maintaining social relationships and by joining social groups. Understanding describes the human need to comprehend the events around oneself, and one's place within social hierarchies and cultures; it is accomplished

by positing explanations for behaviors, making attributions to people or the situation, and even by worshiping deities. Controlling describes the human need to exert influence on one's environment (both social and physical); to act as an agent and to accomplish this by manipulating one's behavior and attempting to manipulate the behavior of others. Enhancing the self describes the human need to feel good about oneself, to have high self-esteem, and is accomplished by maintaining a positive view of oneself and having others view oneself in a similarly positive manner. Trust describes the human need to believe that other people have our best interest at heart. People may often trust other people as a default response because they view the world as benevolent, and violations of trust motivate people to either end the social relationship or to repair the trust (though I question whether this is always the default response in chapter 6).

3. By magical thinking, I do not mean pop cultural definitions of magic; instead, the term is social psychological jargon to describe a phenomenon in which people attribute agency to themselves or others in an attempt to explain (often difficult to explain) outcomes (Pronin, Wegner, McCarthy, & Rodriguez, 2006). Ironically, early studies that used this jargon actually had voodoo dolls serve as props in their experiments.

References

Amodio, D. M., & Frith, C. D. (2006). Meeting of minds: The medial frontal cortex and social cognition. *Nature Reviews. Neuroscience*, 7, 268–277.

Botvinick, M. M., & Cohen, J. D. (2014). The computational and neural basis of cognitive control: Charted territory and new frontiers. *Cognitive Science*, 38, 1249–1285.

Dávid-Barrett, T., & Dunbar, R. I. M. (2013). Processing power limits social group size: computational evidence for the cognitive costs of sociality. In *Proceedings of the Royal Society, B*, 280(1765), 20131151.

Fiske, S. T. (2003). Five core social motives, plus or minus five. In *Motivated Social Perception: The Ontario Symposium* (Vol. 9, pp. 233–246). Mahwah, NJ: Lawrence Erlbaum Associates Inc.

Fiske, S. T., & Taylor, S. E. (1984). *Social cognition reading*. Boston, MA: Addison-Wesley.

Fiske, S. T., & Taylor, S. E. (1991). *Social cognition* (2nd Edition). London, UK: McGraw Hill.

Fiske, S. T., & Taylor, S. E. (2013). *Social cognition: From brains to culture.* Los Angeles, CA: Sage.

Frith, U., & Frith, C. (2001). The biological basis of social interaction. *Current Directions in Psychological Science*, 10, 151–155.

Gallagher, H. L., & Frith, C. D. (2003). Functional imaging of "theory of mind." *Trends in Cognitive Sciences*, 7, 77–83.

Harris, L. T., & Fiske, S. T. (2009). Social neuroscience evidence for dehumanised perception. *European Review of Social Psychology*, 20, 192–231.

Haxby, J. V., Gobbini, M. I., & Montgomery, K. (2004). Spatial and temporal distribution of face and object representations in the human brain. In M. Gazzaniga (Ed.). *The cognitive neurosciences* (pp. 889–904). Cambridge, MA: MIT Press.

Horner, V., & Whiten, A. (2007). Learning from others' mistakes? Limits on understanding a trap-tube task by young chimpanzees (*Pan troglodytes*) and children (*Homo sapiens*). *Journal of Comparative Psychology*, 121, 12.

Humphrey, N. K. (1976). The social function of intellect. In P. P. G. Bateson & R. A. Hinde (Eds.), *Growing points in ethology* (pp. 303–317). Cambridge: Cambridge University Press.

Lee, V. K., & Harris, L. T. (2013). Dehumanized perception: Psychological and neural mechanisms underlying everyday dehumanization. In P. Bain, J. Vaes, & J.-P. Leyens (Eds.). *Are we all human? Advances in understanding humanness and dehumanization*. New York: Psychology Press.

Lee, V. K., & Harris, L. T. (2014a). A social perspective on debates about free will. In W. Sinnott-Armstrong (Ed.). *Moral Philosophy: Free Will and Moral Responsibility* (pp. 381–396). Cambridge, MA: MIT Press.

Lee, V. K., & Harris, L. T. (2014b). Sticking with the nice guy: Trait warmth information impairs learning and modulates person perception brain network activity. *Cognitive, Affective & Behavioral Neuroscience,* 14, 1420–1437.

Miller, G. A. (1956). The magical number seven, plus or minus two: Some limits on our capacity for processing information. *Psychological Review,* 63, 81.

Pronin, E., Wegner, D. M., McCarthy, K., & Rodriguez, S. (2006). Everyday magical powers: the role of apparent mental causation in the overestimation of personal influence. *Journal of Personality and Social Psychology,* 91(2), 218.

Seed, A., & Byrne, R. (2010). Animal tool-use. *Current biology,* 20(23), R1032-R1039.

Stiller, J., & Dunbar, R. I. (2007). Perspective-taking and memory capacity predict social network size. *Social Networks,* 29, 93–104.

van Overwalle, F. (2009). Social cognition and the brain: A meta-analysis. *Human Brain Mapping,* 30, 829–858.

Waytz, A., Epley, N., & Cacioppo, J. T. (2010). Social cognition unbound insights into anthropomorphism and dehumanization. *Current Directions in Psychological Science,* 19, 58–62.

1 Tension

A woman is captured by some people; she initially struggles, resisting the loss of her freedom, before succumbing to the raw power of her capturers. She now stands on display for sale in a foreign land. More people inspect her body, squeeze her breasts, lift her arms to get a good look, stroke her soft skin, and play with her hair. There is a din in the market as they discuss her value. The bidding then begins. Finally a winner celebrates, having outbid his competitors. The woman is taken, arms restrained, off to her new reality of servitude and sexual exploitation.

A fruit is plucked from a tree; it gives some resistance before finally succumbing to the downward force created by the hand pulling it off its branch. It now stands on display for sale in a foreign land. People inspect the fruit, gently squeeze it, stroke the skin, and toss it playfully in the air. There is a din in the market as they debate its value. Three for $5 seems a bit harsh. Finally, the customer negotiates a better price, having worn down the vendor. The fruit is taken and placed in a dark, brown paper bag, off to be consumed.

Scenarios like the ones described above occur on countless occasions throughout the world, have occurred at countless times through human history, and continue to occur today. Interestingly, the psychology present in both the capturers and the customers may be remarkably similar; in both cases, the mental state of the woman and the fruit is not considered. In the case of the fruit, this is rational because pieces of fruit do not have mental states, whereas in the case of the woman, the described behavior may lead to horror on behalf of the reader if he or she imagined her mental state (since women do have minds). How must she feel? What were those animals bidding on her *thinking*? My short answer is that the people who capture, buy, and sell the woman are *not thinking* about her mental state, at least not to the same extent that they may think about other people's mental states. And the use of the word *animals* in the final question is intentional to influence your thinking about those people's minds: they certainly must not be human in quite the right way. Ironically, however, this book argues that they are all too human.

Human beings are one of the few species who are capable of and frequently commit both antisocial and prosocial behaviors. But what is it about humans—the evolved social animals—that causes them to have foes, to befriend strangers, and to bring harm to their foes while helping their friends? The psychological concepts *friend* and *foe* depend on the cognitive-affective appraisal of other people, often based on their perceived role (among other situational factors such as the physical environment) in the social environment and on the person holding the concept. Mental state inferences—deductions about the content of another person's mind—generate such appraisals and are guided by social cues both internal and external to the individual (Fiske & Taylor, 2013). Therefore, the social context shapes mental state inferences, a form of social cognition. This book is about flexible social cognition, that is, whether people engage social cognition or not. The central hypothesis states that *a human/nonhuman appraisal results from flexible social cognition.* This human/nonhuman appraisal perhaps promoted survival during a period when the social environment of human beings consisted of small, relatively insular social groups. Specifically, social cognition engagement entails costs and benefits for both the perceiver and the target, resulting in a need for its flexible employment. Although this ability may have proved useful in the ancestral context, its continued use today in modern societies consisting of many, much larger, more integrated and interdependent groups, may lead to the type of social behavior described at the beginning of this chapter.

The reader may now wonder how I define mental state inferences and, indeed, social cognition. *Social cognition* is a deduction that results from hypothesis testing and includes more conscious, complex inferences such as personality assessment, and more unconscious, spontaneous behavior such as mimicry and contagion effects. The former psychological phenomenon is a form of social cognition in that it provides information, via a prior, about the possible current mental state of a person. The latter two psychological phenomena are forms of social cognition because the hypotheses tested all operate in a Bayesian manner via the assignment of weights. Indeed, perceptual processes can also be thought of as involving a Bayesian process in which prior probabilities guide conclusions about the specific object that is seen, smelled, touched, heard, or tasted.

A Metaphor

To assist in clarifying my definition, a metaphor illustrating social cognition engagement is helpful. Imagine a woman fishing in solace on a lake. The fishing rod is extended over the side of the boat, and a baited fishing line extends from the rod into the water. The fishing line is analogous to the appraisal that determines whether social cognition is necessary; that is, the woman uses the fishing line to detect whether she needs to reel in a fish. As the woman continues to sit throughout the day, she may feel movement on the line. This movement may be due to fish nibbling at the bait, underwater currents that tug the line or cause submerged objects to collide with the line, or a fish actually biting the bait and attempting to swim away. The desired source of movement is the fish biting and then attempting to swim away, and when movement corresponding to this scenario is felt along the fishing line, the woman may infer that a fish is hooked and begin to reel in the line (engage social cognition). However, these other sources of movement may also cause her to act, though in different ways. For instance, a gentle tug on the line from a stronger current may wake her out of her doze. People go through their social environment, and from time to time there is a tug on their line—a tug at their social cognition ability. When this happens, they decide (most likely unconsciously), just like the woman fishing, whether (social) emotion, cognition, and behavior are necessary and, if so, what type of action is required, such as exchanging a pleasant greeting, jumping into the air for joy, or turning and running away. We can categorize the range of possible thoughts, feelings, and actions—the behavioral repertoire—as social or not, just as we could categorize the woman's behavioral repertoire in the boat as reeling in a fish or not.

The reader may wonder whether social cognition is the default state. Activation of the default-state network in cognitive neuroscience involves brain regions primarily involved in social cognition, including the medial prefrontal cortex (MPFC), the precuneus, the temporoparietal junction (TPJ), posterior cingulate, and temporal lobes (see Gusnard & Raichle, 2001; Mars, Neubert, Noonan, Sallet, Toni, & Rushworth, 2012, for reviews).

Indeed, one can argue that what people are doing when they are given no psychological or behavioral task (such as simply lying in the magnetic resonance imaging (MRI) scanner, commonly how the default mode is measured) is thinking about social relationships while mind wandering. Thus, why is the tug on the line necessary? However, people do not spend their days lying in a dark, loud scanner without a task: they are constantly navigating both a physical and a social world. This will at times require social cognition and at times not. Therefore, the term default state is a misnomer and is relative only to cognitive neuroscience experiments, not people's everyday or default experiences in the real world.

One possible objection to this metaphor wonders why an appraisal process needs to be the criterion for engaging social cognition. Why not some perceptual ability? Social cognition arises in part from perceptual feature space mapping just as do other perceptual detections such as visual distinctions that cause a person to identify a yellow car in New York City as a taxi, versus any other-colored car. The difference is that a human/nonhuman distinction is not based solely on perception because person perception is a two-part process: a physical perception of the other person's physical form and a mental state inference about the contents of the person's mind. Therefore, social cognition requires a trigger (present, implied, or imagined) to engage. Here, I argue that visual features are not necessary or sufficient because social cognitions are largely inferences, not perceptions. Just like the cause of the fishing line's movement below the surface of the water, social cognition is about the unseen, the inferred, the appraised. Therefore, it is deserving of its own metaphor separate from vision. Second, just as with yellow cars in New York City, the social environment determines whether social cognition is necessary or not. If a person were in Washington, DC, where the cabs come in all colors, other visual distinctions would be necessary, such as whether the car had advertising on its roof. Similarly, if the lake on which the woman fished experienced turbulent weather, the same movement on the fishing line that elicited reeling on a calm day would lead to no such result on this windy day. Therefore, social cognition engagement is inherently flexible.

Flexibility

I describe social cognition use as *flexible;* flexibility is herein defined as the property of a physical object or psychological phenomenon to stretch during use, and the ability of a person using a physical object or psychological phenomenon to withhold its use. This is not an English dictionary definition of flexibility—though it captures the ability to change or be changed sentiment that is inherent in such a definition—but jargon that permits the encompassing of two related yet distinct ideas: *stretching* and *withholding*. Flexibility makes physical objects and psychological phenomena adaptable in a variety of contexts; an inflexible or rigid object or phenomenon is suitable only for the select contexts where the properties of that object or phenomenon are useful, and an object or phenomenon that cannot be withheld and must be used in all contexts has compromised efficiency. Therefore flexibility refers to both a property of the object, fungibility, and a property of the user of the object, restraint.

Stated differently, flexibility ensures that the *tension* applied to an object or phenomenon does not allow it to succumb, rendering it useless. Flexibility preserves the object or phenomenon when tension is deemed too great. *Tension* is a force applied to an object or phenomenon and may be considered a demand on said object or phenomenon to perform its intended function the right way. In the case of a physical object such as a rope, the object applying the force to the rope supplies the demand, and the intended function of the rope is to hold the object. Thus, for physical objects, tension (P_t) is exhibited by the weight or mass (m) of the object supplying the demand, multiplied by acceleration or gravity[1] (g) present in the context

$$P_t = m \times g$$

In the case of a psychological phenomenon such as social cognition, the agent requiring evaluation supplies the demand, and the intended function of social cognition is to make an accurate evaluation about that agent. Thus, for mental constructs, tension (M_t) is exhibited by the attributes (a) of the agent supplying the demand combined with the social contextual (s) factors present

$$M_t = f\,(a, s)$$

This simple equation[2] captures a basic tenet of social psychological theory: person–situation interaction results behavior.

I should note immediately that this equation is no different from the classic equation first outlined by Kurt Lewin (1935), an equation that underlies all theory in social psychology:

$$B = f\,(P, E)$$

In this formulation, B represents a person's behavior, and P and E capture personality and environment, respectively. Personality refers to causal factors for behavior internal to the person (a in the mental tension equation), whereas environment refers to causal factors external to the person (s in the mental tension equation). Therefore, all I have done is to recast an old interaction term initially used to explain behavior for mental tension. By doing so, I have used the interaction term to explain a *precursor to behavior,* not behavior itself. Hence, in Lewin's formulation,

$$B = f(P, E) + e$$

or

$$B = M_t + e$$

where e is an error term (unexplained variance). Thus, the current formula simply adds error to the function in Lewin's formulation instead of hiding the error within either the person or the situation.

A mathematician may wonder whether the variables just described (a and s) are actual numbers, making multiplication or any other mathematical function possible. I argue that they are (particularly in the brain), with a range of 0 to 1, where 0 implies no impact of the variable and 1 implies complete impact. However, in reality, I cannot imagine a situation where either variable would be 0, since for every behavior both the person and the social context have an impact on behavior. Similarly, it is difficult to imagine a scenario where the person or social context has absolute or complete impact on behavior, making 1 highly unlikely. But there are cases where the person or social context can have more or less impact on behavior. Additionally, their relationship is not hydraulic, that is, the person and the social context can both have large or small impacts on behavior, with the error term capturing a lot or very little unexplained variance for the behavior.

But how does stretching and withholding benefit social cognition? Flexibility can be an asset for any object or phenomenon. Let us first consider the case for stretching. The ability of physical objects such as ropes to stretch ensures that they do not succumb as easily to increased tension; the more ropes are able to stretch, the less likely they are to break and fail to perform their function the right way. Similarly, stretching social cognition prolongs its usefulness, making its use applicable in a variety of contexts (such as toward nonhumans). If social cognition could only be used in a narrow context (such as only when interacting with family members), then this phenomenon is limited, and people would require a host of other psychological phenomena for each unique context. The ability of social cognition to stretch ensures that it does not succumb as easily to increased demand; the more social cognition is able to stretch, the less likely it is to lead to incorrect conclusions over time and fail to perform its function the right way, becoming useless. This is because increased use in a variety of contexts improves social cognition performance as the person gains experience or learns to use his or her social cognition appropriately across a larger sample of contexts. For instance, stretching social cognition to the wind rattling the windows may cause me to think that a burglar is attempting to enter the house. However, this experience and the falsification of the burglar theory lead to more efficient use of social cognition in the future when the wind rattles leaves, hedges, or roof tiles.

The case for withholding is slightly less complex. The ability to withhold the use of a physical object like a rope also preserves its longevity. For instance, if the tension exerted on ropes is expected to be too great, using other objects such as chains preserves the use of ropes for more hospitable occasions while it still allows the completion of the task required in the context. Similarly, withholding social cognition preserves its usefulness (I describe my own research effort on dehumanized perception that establishes how this is possible in greater detail below). For instance, if the tension exerted on social cognitive abilities is expected to be too great, using another mental construct such as an affective appraisal preserves the validity of social cognition for a more hospitable occasion where it is more likely to lead to the right conclusions and/or less cognitive load while it allows the completion of the task required in the context.

Finally, it is important to note that I am agnostic about motivations to utilize stretching or withholding of social cognition. Stretching social cognition does not depend on motivations to engage the process, and neither does withholding. However, the case for withholding may require a bit more of an argument. People can withhold use of a psychological phenomenon because it is not necessary: one may withhold use of logical problem solving when watching clouds cascade over the horizon because logical processing is not necessary in that instance (unless one is a meteorologist). Similarly, one may withhold use of social cognition in a particular instance because its use is deemed superfluous. Intrinsically, both features of flexibility do not require a motivational component.

Mental State Inferences

The definition of mental state inference is crucial to the theory I propose below. I view mental state inferences as a form of social cognition: a way of figuring out what is in someone else's head. Sadly, the term "mental state inference" is conflated in the psychology and neuroscience literatures. Here, I define *mental state inference* as the process by which people deduce the thoughts, feelings, dispositions, and intentions of other agents. I further argue that this process is Bayesian, inferring the likelihood of specific mental states based on priors and evidence. Mental state inferences are first and foremost *inferences*—logical conclusions based on evidence. For instance, given:

a

If *a*, then *b*

∴ *b*

This simple formal argument is an example of the logical process at the root of most inferences. It should be noted that philosophers have identified a flaw with such arguments: they are only valid if *a* is assumed to be true (see for instance Harman & Kulkarni, 2012). Nonetheless, the variables *a* and *b* can represent both person and social context information. However, a mental state inference, and indeed social cognition, is not simply an inference or cognitive process applied to people; what is remarkable is that mental state inferences and broader social cognitions consider priors from behavior, relying on statistical information about people. Philosopher and social psychologist Fritz Heider (1958) referred to people as naive scientists because of their integration of statistical information and hypothesis testing. Social psychology carefully describes this Bayesian process since it remains a key method used during social cognition. Statistical updating itself relies on logical reasoning, so mental state inferences are rooted in logic. This point is important because people often *seem* to be fundamentally irrational in behavior resulting from social cognition. I hope to return some rationality to people's behavior by highlighting this logical process.

A critic may wonder whether unconscious social cognitive processes such as emotion contagion and mimicry also constitute mental state inferences as defined in such a Bayesian manner; surely consciousness is a prerequisite for mental state inferences. I argue that a Bayesian process may also account for such unconscious occurrences of social cognition via the assignment of weights to the priors. All social cognition is hypothesis testing, whether conscious or not, but some forms of social cognition operate based on differential weighting of priors. For instance, facial expression mimicry requires a perceiver to recognize another's facial expression, then to execute that expression herself. Researchers have identified mirror neurons in motor cortex that fire both when one is observing behavior and when one is performing the same behavior (see Rizzolatti & Craighero, 2004, for a review). This has been posited as a possible mechanism for mimicry. However, appeals to mirror neurons are not necessary to explain such behavior: a simple changing of weighting functions can explain how such fairly spontaneous processes occur.

Next, I dissociate mental state inferences from other social cognitive terms commonly used in the literature such as empathy, perspective-taking, emotional contagion, theory of mind, and mind perception. Empathy requires a mental state inference (or alternatively emotional contagion) but is not the same as a mental state inference (see Batson, 2009, for a similar discussion). Empathy requires feeling what another person feels and then being motivated to help that person (Batson, 1991). Additionally, this process may or may not be conscious (Batson, 2015). Empathy is usually described in situations of distress or other negative feelings. A desire to relieve the negative feeling in either the target or the perceiver results in the motivation to help. Therefore, a mental state inference is necessary to trigger empathy but is not itself an empathic response. For instance, I can infer that the attractive person on the train is probably unintelligent because of stereotypes I may hold about attractive people. I can also infer that the person is bored, restless, and maybe even impatient as the train tugs along by reading her body language but also by integrating knowledge about what people are generally like on a long train ride, as well as maybe integrating my own internal states (I may also be bored and restless, and I assume everyone else is as well). At no point in any of these trait or situational inferences is empathy present. It may be triggered as I arrive at my stop, leaving the attractive person back on the train, motivating me to remark to her on the tiresome nature of trains as I pass by, bringing a brief smile and temporary relief to that person's dour experience. But this empathic response first depends on the mental state inference that the person is bored and restless. Therefore, empathy and mental state inferences are not synonymous.

Perspective-taking perhaps comes the closest to mental state inferences as a form of social cognition. Perspective-taking entails putting oneself in another person's shoes, that is, experiencing a situation from another person's point of view (Galinsky & Moskowitz, 2000). Again, a mental state inference is necessary to deduce the context of the other person's mental state but is not itself perspective-taking; assuming another's point of view as one's own point of view is needed to complete the phenomenon. Worded differently, a mental state inference is necessary for perspective-taking—but not sufficient. For instance, I can take the perspective of someone on a long train ride, but I first have to imagine what her mental life in that situation is like. Imaging such a mental life requires mental state inferences but is not itself perspective-taking. The specific individual to whom this inference is applied (Jack, the experienced commuter, versus Jill, the first-time train passenger) flavors the mental state inference to become another person's perspective, adding person-specific information. Jack's perspective will involve calm, ease, relaxation, a feeling of familiarity with the experience, and inferences of lots of mundane thoughts. In contrast, Jill's experience will also be consistent with the inference, but it may involve far more rumination about the experience, a feeling of unease or excitement, impatience, and perhaps a sense of wonder as the landscape goes by. The initial mental state inference about train passengers delivers social information, activating a number of possible mental states, whereas the specific person in the situation reduces that number. Finally, as the perceiver, I must respond from either from Jack's or Jill's perspective, ignoring my own beliefs, attitudes, and biases about train rides or train passengers.

Emotional contagion describes feeling the emotion that another person is feeling (Hatfield, Cacioppo, & Rapson, 1994). This is often an unconscious process, and researchers describe emotional contagion to be similar to contracting a virus: the observer "catches" the emotion of the other person. This may occur via spontaneous mimicry processes (this work actually grew out of many studies utilizing facial electromyography [EMG] to study spontaneous mimicry), which simulate the emotional experience of the other by causing the observer's own physical display of emotion to mirror that of the other. For instance, when one observes another smiling because she is happy, the observer subtly smiles herself. Presumably, the brain of the observer is able to obtain information from the activation of the facial muscles involved in creating the smile, causing the brain to determine that the observer must herself be happy (although I argue that situational information must also be integrated in order to result in the emotion of happiness, since smiling could occur because the person is deriding the observer, in which case happiness would be an inappropriate emotional response). Mimicry is a form of social cognition, and one may consider emotional contagion an affective version of mental state inferences because integration of situational information is necessary in order for contagion to occur (although I hesitate to differentiate psychological processes into affective and cognitive camps since most affective processes require cognition and most cognitive processes trigger affect).

Theory of mind is borne out of the same intellectual tradition that I review in the coming chapters: it is a tradition that views people as naive scientists and social cognition to be the result of statistical, scientific inferences. However, the term "theory of mind" has often been used in the literature to characterize much more complex phenomena than simply mental state inferences. As an example, a popular test of theory of mind abilities uses a false belief task (Baron-Cohen, Leslie, & Frith, 1985). In this task the observer should not only infer a single agent's mental contents but have access to whether the agent believes those contents are true or false. This complex *metacognition* goes beyond simple mental state inferences as defined above. Metacognition is sometimes sufficient for a mental state inference, but it is not necessary. In fact, the classic Sally–Anne false belief task involves a number of mental state inferences that constitute the metacognition necessary to solve the task. The Sally–Anne task involves presenting the subject with a scenario depicting two agents, Sally and Anne. Both agents are in a room (situation) with two opaque boxes. Sally has a ball that she puts in the first box. Sally then leaves the room. While Sally is gone, Anne takes the ball from the box and places it in the second box. The question posed to the subject is when Sally returns, where will she look for the ball? To correctly solve the task and report "the first box," the subject must understand that Sally will look for the ball where she left it. That means the subject, even though aware of the true location of the ball (cognition), ignores this information (inhibition of cognition) and responds to the task from Sally's perspective (results mental state inference). How does this form of social cognition finally occur? By enacting the following logical (if-then) argument, we have: if Sally were not aware of the switch, then she should act like an agent who does not know the true location of the ball (prescriptive norm). It should be noted that Sally's behavior is not random; Sally should look for the ball where her prior information indicates the ball is located, that is, in the first box. Therefore, my objection to the term "theory of mind" is with the tasks used to test the psychological phenomenon, not the use of the term itself.

Finally, mind perception is a bit of a misnomer since the mind is abstract and cannot be perceived; mental states can only be inferred (I admit I use the term "mind" in this book since I am writing for a diverse audience). In essence, descriptions of mind perception are confusing in the literature because this term does not properly characterize the psychological phenomenon. The term has sometimes been used to describe mental state inferences (Waytz, Gray, Epley, & Wegner, 2010), but not exclusively. For the purpose of this book, if I discuss a phenomenon that falls under the rubric of mind perception, I will clarify precisely the social cognition process present in the research. At best, mind perception is useful as a lay term to capture folk intuition regarding social cognition. But as scientific jargon, it fails to accurately characterize the phenomenon. A mind perception champion may argue that the term has been used to describe inferences about the quality and capacities of other minds, whereas mental state inference captures inferences about the content of those minds. However, this champion would still rely on the term "inference," making the use of the word "perception" a misnomer.

Nonetheless, there are some philosophers who believe that minds can be directly perceived. Proponents of the phenomenological proposal argue that we embody mental states, allowing us to directly perceive them (see Gallagher & Zahavi, 2013). This approach denies the principle of imperceptibility on which both theory-theory and simulation theory accounts of mental states are based. Stated differently, the principle of imperceptibility suggests that mental states are hidden and therefore must be inferred via some mechanism: either simulation or theory testing. The argument against this principle suggests that we have bodily reactions to other people's mental states that we can perceive via phenomenology (experience). This personal experience with other mental states is a direct perception similar to other personal experiences we have with other entities that we directly perceive with vision, audition, olfaction, taste, or touch. Each of these sensory experiences leads to bodily changes that we directly perceive and experience, and mental states are no different. Such a theory supports the term mind perception. However, I argue that such a phenomenological account still requires an inference because we have to somehow interpret bodily responses. This line of argument, if pushed further, would suggest then that nothing is directly perceived, only inferred from bodily responses. Certainly, psychological evidence such as the McGurk effect (McGurk & MacDonald, 1976) that I describe below brings this into question, but this debate is by no means resolved.

Therefore, the social cognition terminology across and within the various disciplines I expect to cover is confused. However, there is significant overlap among the variety of terminology with the definition of mental state inferences I have outlined here. Because of this overlap of terminology in the literature, I only describe phenomena across the various literatures that fit my definition and not those used by the authors of these empirical studies. Stated differently, I describe phenomena consistent with my definition that may have been mislabeled in the literatures.

Social Groups

Finally, a brief word about social groups—two or more individuals who share a common identity[3] (Fiske, 1992; Hogg, 2014; Levine & Moreland, 2012). This definition comes directly from social psychological and sociological theory and hinges on the phrase *common identity*. An identity is defined as a mental construct representative of a type of person. This mental construct or *schema* captures the prototypical person subsumed by the identity, making accessible physical and psychological features of such a person. Identities also describe relationships between that prototypical person and other people with different, relevant identities. Intrinsically, then, identities become supremely important for intergroup relationships because they help define group boundaries. At this point, it behooves me to clarify that identities can be real or constructed. A realistic view of identity is age: old or young. People adopt such an identity and remain in that identity until they transition from one age range to another. Such identities are conferred by nature and people's biology,[4] not societal or self-prescriptions. Constructed identities are categories such as race or ethnicity. People confer such identities to themselves, making such categories malleable. The recent case of Rachel Dolezal and historic cases of racial "passing" of African Americans as Caucasian Americans illustrate the malleable, nonpermanent nature of socially constructed identities. These latter types of identities are primarily my focus in this book.

People have many identities, each becoming salient in different social contexts. Such contexts can be large, such as being aware that one is in London, or more narrow, such as being aware that one is sitting in a Greek restaurant in a popular, tourist-filled neighborhood in London. This difference between types of contexts facilitates identity salience. For instance, someone's identity as a husband is more salient at home than it is at the office (unless that person works as a marriage counselor perhaps). This feature of people's identity is unique to human beings. Nonetheless, some identities are more readily accessible or chronically salient; these include gender, age, race, ethnicity, socioeconomic, national, and religious identities. However, this is a fairly recent development in human social life—hunter-gatherer humans belonged to a small number of primary groups (tribe, age, gender, and a few others perhaps), not many. As a result, saliency of identity would have been less important, and a simple us-versus-them distinction should have been sufficient.

Here, I describe a determination of whether social cognition should be engaged or not because of an impending interaction with a person who is human or not and thus deserving of social behavior or not. Thus far, I have spoken about mental state inferences, flexibility, and social groups. I introduced the central concept of tension, and I intend to argue that flexible social cognition results from the development and arrangement of boundaries and conditions necessary for the existence of social groups—that is, criteria for group inclusion and exclusion. People stretch and withhold social cognition when the social context encourages such behavior. Having defined the key concepts, I then discuss possible social contexts that may have shaped social cognitive abilities and properties. Herein begins the discussion of the interaction of the person and the social context; it emerges from the conversation about why distinctions between potentially benevolent (helpers) and malevolent agents (harmers) are necessary. But first, I explore the context during evolution in an attempt to understand how and why social cognition may have become flexible.

Notes

1. It should be noted that one of Albert Einstein's greatest accomplishments was to dismiss gravity as a force exerting influence of objects. However, since I am not a physicist, the metaphor is sufficient for my purposes.

2. In this equation, the relationship between *a* and *s* is described using a function, rather than multiplication as is found in the physics equation, because the precise relationship between the two variables is unknown. For instance, the impact of the situation could be additive to the agent, or the interaction between the agent and situation, or agent divided by or subtracted from the situation; any of these possibilities could produce the mental tension. Therefore, the function allows for many possible types of relationships between the variables, which can be further specified for mathematical computations.

3. At least one philosopher, reading earlier drafts of this book, has wondered whether individuals with a common identity must know each other. For instance, do twins who were separated at birth comprise a social group? My answer is no because a common identity is based on a shared mental representation. Therefore, siblings comprise a social group because they both view themselves as having the same parents and will identify as such. Separated twins who discover each other may

become a social group once the shared mental representation is established, but they do not comprise a social group until such time.

4. It should be noted that people often attempt to change their age identity by manipulating their appearance and behaviors. However, biologically, an eighteen-year-old and eighty-year-old are markedly different, despite attempts at identity mutability.

References

Baron-Cohen, S., Leslie, A. M., & Frith, U. (1985). Does the autistic child have a "theory of mind"? *Cognition*, 21, 37–46.

Batson, C. D. (1991). *The altruism question: Toward a social-psychological answer*. Hillsdale, NJ: Lawrence Erlbaum Associates.

Batson, C. D. (2009). These things called empathy: Eight related but distinct phenomena. In J. Decety & W. Ickes (Eds.), *The social neuroscience of empathy* (pp. 3–15). Cambridge, MA: MIT Press.

Batson, C. D. (2015). *What's wrong with morality? A social-psychological perspective*. New York: Oxford University Press.

Fiske, A. P. (1992). The four elementary forms of sociality: Framework for a unified theory of social relations. *Psychological Review*, 99(4), 689–723. doi:10.1037/0033-295X.99.4.689.

Fiske, S. T., & Taylor, S. E. (2013). *Social cognition: From brains to culture* (2nd ed.). Los Angeles, CA: Sage.

Galinsky, A. D., & Moskowitz, G. B. (2000). Perspective-taking: Decreasing stereotype expression, stereotype accessibility, and in-group favoritism. *Journal of Personality and Social Psychology*, 78(4), 708–724. doi:10.1037/0022-3514.78.4.708.

Gallagher, S., & Zahavi, D. (2013). *The phenomenological mind*. New York: Routledge.

Gusnard, D. A., & Raichle, M. E. (2001). Searching for a baseline: Functional imaging and the resting human brain. *Nature Reviews. Neuroscience*, 2(10), 685–694.

Harman, G., & Kulkarni, S. (2012). Statistical learning theory and induction. In N. M. Seel (Ed.). *Encyclopedia of the sciences of learning* (pp. 3186–3188). New York: Springer.

Hatfield, E., Cacioppo, J. T., & Rapson, R. L. (1994). *Emotional contagion.* Cambridge, UK: Cambridge University Press.

Heider, F. (1958). *The psychology of interpersonal relations.* Hoboken, NJ: John Wiley & Sons. doi:10.1037/10628-004.

Hogg, M. A. (2014). From uncertainty to extremism: Social categorization and identity processes. *Current Directions in Psychological Science,* 23(5), 338–342. doi:10.1177/0963721414540168.

Levine, J. M., & Moreland, R. L. (2012). A history of small group research. In A. W. Kruglanski, W. Stroebe, A. W. Kruglanski, & W. Stroebe (Eds.). *Handbook of the history of social psychology* (pp. 383–405). New York: Psychology Press.

Lewin, K. (1935). *A dynamic theory of personality.* New York: McGraw-Hill.

Mars, R. B., Neubert, F. X., Noonan, M. P., Sallet, J., Toni, I., & Rushworth, M. F. (2012). On the relationship between the "default mode network" and the "social brain." *Frontiers in Human Neuroscience,* 6, 189.

McGurk, H., & MacDonald, J. (1976). Hearing lips and seeing voices. *Nature,* 264, 746–748.

Rizzolatti, G., & Craighero, L. (2004). The mirror-neuron system. *Annual Review of Neuroscience,* 27, 169–192.

Waytz, A., Gray, K., Epley, N., & Wegner, D. M. (2010). Causes and consequences of mind perception. *Trends in Cognitive Sciences, 14*(8), 383–388. doi:10.1016/j.tics 2010.05.006.

2 Evolution

A bonobo named Lola walks into a testing chamber. A pile of food sits in the middle of the chamber, equidistant from her entrance and a second entrance at the other side of the chamber. Behind this second entrance sits Elo, an unfamiliar bonobo. The food is positioned such that Elo cannot reach it. There are two ropes leading away from each entrance. Lola has previously learned that pulling the rope can open the entrance, permitting passage in and out of the food chamber. She has the option of (1) pulling the rope leading from Elo's chamber, opening that entrance, which would allow Elo to enter the food chamber and share the food, (2) pulling the other rope, which opens the chamber from which she came, or (3) pulling neither. After inspecting the food but before taking a bite, Lola gives a sharp tug on the first rope, opening Elo's door, allowing him to come in and share the food.

A group of five male chimpanzees patrol their territory in the Congolese forest. The male at the head of the group hears the sound of another chimpanzee. Quickly, all five race to the edge of their territory on the banks of the river. There, they find an unknown male chimpanzee alone, foraging for food. In a swift motion, all five descend upon the loner, viciously ripping him to shreds.

Human beings are not the only social species capable of pro- or antisocial behavior (Kling, 1975) or social cognition (Hare & Tomasello, 2005). The examples above depict the behavior of two different, yet closely related, species. Evolutionary anthropologists often use such contrasting examples to illustrate how small genetic variance can lead to very different types of behavior. However, one should note that bonobos are not always prosocial, and chimpanzees can exhibit prosocial behavior. Nonetheless, the contrasting examples highlight an important feature of evolution: the social context can lead to very different types of behavior. Bonobos evolved in an environment with much less food competition, whereas chimpanzees evolved in an environment where food competition was much higher and included competition from other species in addition to other chimpanzees (see chapter 7 for further explanation). There are other important differences between bonobo society and chimpanzee society that I describe below.

But first, it is important to point out why cross-species comparisons are necessary. The defining feature of *Homo sapiens'* dissociation from other apes, particularly our closest genetic relatives the chimpanzees and bonobos, surrounds the more complex mental life of human beings. These advanced problem-solving skills and the capacity for abstract thought have pushed humans beyond other species. Therefore, probing these capacities in our genetic relatives allows an understanding of the specific qualities of human experience that are unique.

Although debates continue to rage over whether other primates experience consciousness (see Mashour & Alkire, 2013, for recent review), there is no debate as to whether they are social beings; like humans, they accomplish the basic skills survival and reproduction in a communal fashion. Since nonhuman primates share a large genetic similarity to humans, suggesting a relatively recent common ancestor, the existence of shared traits that promote survival and reproduction suggests that this ancestor possessed these traits. The interaction between a common ancestor and the physical and psychological–social challenges of the time perhaps promoted specific traits as advantageous, ensuring our very existence today (Humphrey, 1976).

In contrast, comparisons need not only be made based on the genetic raw material shared by a common ancestor, material necessary for behavior. Other species have a variety of methods for accomplishing these basic tasks facing existence on planet Earth, hinting at approaches shaped by the demands of life on the planet. Comparisons of other species' survival in similar physical and social environments suggest a separate contribution relatively independent of genetic raw material—an environmental account. This nature-nurture debate, although never completely solved, is irrelevant because these comparisons can simply imply relative trait contributions from both the environment and genetics, with the truth lying somewhere in between (or in the interaction term). Therefore, to examine uniquely human traits such as flexible social cognition is to examine an interaction effect between environment and genetics.

The two basic tasks for species existence are survival and reproduction (Darwin, 1859). The relationship between them is not mutually exclusive because survival without reproduction would not ensure the continuation of the species' existence (or at least one's own genetic lineage), and there is no chance at reproduction if the organism does not survive to sexual maturity and beyond. As a result, survival supports reproduction—survive long enough to reproduce healthy offspring (although evolutionary anthropologist colleagues often tell me healthy grandchildren). Survival and reproduction depend on the emotions fear and reward, respectively. Reward is considered in this context positive feedback, almost always accompanied by positive affect—the amount of conscious feeling and/or dopaminurgic release in the striatum associated with the experience of the emotion—no matter how minuscule. Fear, conversely, operates through relatively large levels of negative affect. This carrot-and-stick reinforcement facilitates accomplishment of basic survival tasks, and a cross-species comparison demonstrates how these two emotions are shared and divergent in humans and other species. We discuss each emotion in more detail when discussing prediction in the next chapter.

Genetics × Environment: An Example

At this point we can turn to genetics, but perhaps it may be better to begin our discussion with a consideration of environment, specifically the social environment. A description of the social environment requires a description of the physical environment to account for the interaction between geography and psychology. The physical environment is relevant because it has always determined the extent to which human beings could have physical and social interaction, thus shaping the tasks necessary to accomplish survival and reproduction. This has been particularly true for virtually all of human evolution, although recent technologies have facilitated social interaction between and among people across large distances (I discuss the possible consequences of this at the end of the final chapter). However, the Internet, telephones, and the printing press (which facilitated written communication over large distances) are relatively recent inventions in human history, and our brains' biology and psychology essentially evolved within limitations on human social interaction.

For instance, let us assume all human beings coming out of Africa 600,000 years ago—whose descendants comprise modern human beings—had a trait that promoted either a one male–one female reproductive unit or promiscuous sexual behavior. Although humans have more pair-bondings (i.e., one male–one female reproductive units) than other primates, humans are also promiscuous. Therefore, they can (1) practice "sequential" pair-bondings or (2) maintain multiple sexual partners at once. The former is practiced by many modern hunter-gatherers—a male and a female bond until their offspring weans—then they break up the bond and move on to other partners (for a genetic account of this behavior, see Walker, Flinn, Hill, & Ellsworth, 2011). Such relationships last approximately seven years (Gray & Anderson, 2010).

Other mammals also engage in pair-bonding, such as prairie voles. These mammals also practice sequential pair-bonding; they find a mate at puberty and co-reside until they have children who have weaned. Other researchers have identified *oxytocin* and *vasopressin* as hormones that underlie such behavior (Carter, 1998; Ross, Cole, Smith, Neumann, Landgraf, Murphy, et al., 2009; Ross & Young, 2009; Williams, Insel, Harbaugh, & Carter, 1994; Young, Lim, Gingrich, & Insel, 2001). *Dopamine* may also be involved in pair-bonding behavior; specifically, manipulating dopamine D2 receptors results in prairie voles that change their mating preferences, impairing partner preferences (Gingrich, Liu, Cascio, Wang, & Insel, 2000; Wang, Yu, Cascio, Liu, Gingrich, & Insel, 1999). Given that such hormonal effects may also drive pair-bonding in humans (see Churchland & Winkielman, 2012 for a discussion), this overlap of hormonal effects with other mammals suggests a strong interpretation that such preferences are hard-wired from a common mammalian ancestor.

However, imagine that nuclear family life was reserved for some human beings. In modern descendants of those people who could not have nuclear families, high rates of single-parent and extended parent family households (grandparent living together, with, or in close proximity to their bonded children and grandchildren) with corresponding low rates of nuclear family households can perhaps be attributed to ancestral behavior. Specifically, in these descendants, perhaps the trait(s) that encourages reproduction prioritizes a higher number of sexual partners. Given that sexual relationships, even casual sexual relationships in humans require psychological relationships facilitated by social cognition, resulting in *social relationships*, these partners were probably spread across distance to enhance the likelihood of success. Stated differently, if the mates are farther away, then there is less opportunity for conflict and competition between them. Similarly, in descendants of all the other humans who did not have generations of nonnuclear family experiences, the trait that encouraged reproduction continued to prioritize nuclear family relationships. This hypothesis is testable given modern genetic capabilities (if such groups existed), but results would hinge on how much human history—how many successive generations—need to elapse for the environment to have an impact on genetic raw material, promoting a trait with an alternate method for accomplishing a basic survival skill. Such comparison studies (on a variety of issues beyond reproductive behavior) can test whether environmental impacts on genetics that promote sexual behavior is indeed possible in humans and can help to answer questions relevant to social development.

However, the interesting component of the example above for our purposes is the psychological demand present in the environment that would lead to prioritizing an alternate set of behaviors to accomplish a basic survival skill. In the equation for mental construct tension,

$$M_t = f(a, s)$$

the agent supplying the demand is the other person in the social relationship, and the social contextual factors present reduce freedom for engaging in social relationships. If I set the agent's demand to 1, signifying a positive response and willingness to engage in a sexual relationship, and the social contextual demand to 0 because of an environmental restriction preventing the relationship, then the equation solves to 0. Thus, it suggests that since mental construct tension is required for social functioning; any person would focus his or her energies on changing the value of social contextual demand from 0. Therefore, in the restrictive reproductive context described above, sexual relationships persist, perhaps under the cover of darkness and with many partners over large distances. If the restriction was imposed by society, making it much more absolute, this could have likely involved great risk to life and often resulted in reduced survival, which would determine whether reproduction occurred before survival ended. Presumably, this adaptive behavior would be successful if there existed many descendants from such illicit relationships who displayed the modified trait. This parsimonious explanation of infidelity will never completely explain the phenomenon—other social contextual factors often interact with other genetic traits subserving more proximal psychological and social goals—but it provides a testable hypothesis for scientific research, the goal of a social neuroscience approach.

Basic Emotions

This hypothetical example also highlights the supremacy of reward above fear, loss, and even death, the ultimate failure of survival. People may be willing to risk survival for reproduction with a desirable mate. Reward motivates a vast array of approach behaviors, initiating learning processes to reinforce reward-related behavior, whereas fear monitors the environment for threats, initiating learning processes to avoid fear-inducing behaviors and loss.

Given that there are other basic emotions such as sadness, anger, and disgust (fear and surprise comprise the same facial expression, but usually result in divergent outcomes—experiencing agony and relief from experiencing agony, respectively), one might wonder how they, as a suite of affective processes, promote survival and reproduction. Fear is perhaps the most basic of basic emotions, occurring in response to perceived threats to the self, real or imagined. Fear is necessary for survival because it leads to avoidance behavior. Because of this primary importance, it enhances cognitive abilities including perception and attention (Ferneyhough, Kim, Phelps, & Carrasco, 2013) memory and learning (Dunsmoor, Murty, Davachi, & Phelps, 2015), and decision making (Hartley & Phelps, 2012). It relies on brain regions involved in learning and decision making and directs most of the brain to achieve survival (Whalen & Phelps, 2009). Facial expressions of fear include widening of the eyes so that the whites are displayed and raising of the brows. Fear is often accompanied by freezing behaviors both in humans and other animals and prepares the organism for flight or fight.

Sadness occurs to loss and perhaps reinforces fear-induced learning after negative experiences. It requires the individual to socially withdraw and ruminate (think excessively, often about social relationships and the self [Thomsen, 2006]), perhaps in service of restoring social relationships. Sadness displays involve convulsing the body, and facial displays of sadness involve furrowing the brow, wrinkling the nose, and turning down the corners of the mouth. Crying or tears typically accompany sadness as a social signal to other people that a loss has occurred.

Anger responds to goal blockage and ensures that the person experiencing it continues pursuit of the goal. As a result, this approach emotion is often thought of as negative, leading to negative outcomes for social relationships, but it also signals dominance and status relative to determining the composition of social relationships and even their very existence. However, experiencing anger may also be a positive experience because angry people often report feeling empowered (Carver & Harmon-Jones, 2009). Like fear, anger also involves the widening of the eyes as well as the furrowing of the brow and occasionally the baring of teeth.

Disgust tends to set rules to govern social relationships (Haidt, 2001). It is flexible and often specific to particular physical objects. Because of contamination, disgust is also useful for determining which social relationships should be maintained and which should be avoided (Haidt, Rozin, McCauley, & Imada, 1997). Composed primarily of an avoidance behavioral response, the disgust facial expression is often confused as anger by older children and adults (Pochedly, Widen, & Russell, 2012; Widen & Russell, 2010, 2013).

It should be noted that the concept of basic emotions is fast becoming outdated in the emotion literature. For over a century the debate raged as to whether basic emotions occurred or whether people experienced diffuse physiological reactions and interpreted them as emotions based on language and the social context. For instance, proponents of basic emotion (called the Canon-Baird theory after the two researchers who proposed it in the early twentieth century based on the writings of Darwin) argue that a person walking in the forest who encounters a bear experiences fear that then motivates that person to run away. In opposition, the James-Lange theory (also named after the two researchers who proposed it in the early twentieth century) argues that a person who encounters a bear in the forest runs away and, while running away, notices his behavior and deduces that he must be afraid. This alternate order of the emotion and the behavioral response still has not been proven, although recent brain-imaging evidence suggests that people do interpret their physiological responses, and there is no homologue for basic emotions in the brain (Lindquist, Wager, Kober, Bliss-Moreau, & Feldman-Barrett, 2012). Next, I discuss the how and why of social cognition as it occurs across species.

How? Spontaneous Social Cognition

One interesting starting point for comparing human social cognitive abilities is nonhuman primates. There are a range of possible relevant behaviors, including cooperation and helping behavior (for discussion, see Rosati, Santos, & Hare, 2012; Warneken & Tomasello, 2009), but it may better serve the discussion here to narrow the focus to two behaviors shared with humans: contagious yawning and food sharing. These behaviors capture both a spontaneous, less conscious, more automatic, and a more considered, conscious process. Intrinsically, describing both provides a comparison when contrasting the more and less automatic processes. Beyond apes, other primates such as macaques are aware of and detect changes in their social context instantly (Chang, 2013), suggesting the evolutionary history may be even older than a common ape ancestor. These social perceptions guide motivational processes that release a cascade of neural and hormonal processes that constitute the fabric of human experience. The extent to which these processes trigger "social cognition" is debatable since many researchers argue that social processes are simply co-opted cognitive processes, not specialized processes themselves. Nonetheless, our discussion of social cognition includes food-sharing behavior as well as behavioral mimicry including contagious yawning.

Contagious yawning is considered a form of social cognition due to either nonconscious mimicry or to a contagion effect (Yoon & Tennie, 2010). In humans, researchers have demonstrated that contagious yawning relates to self-perception and identification of social faux pas, not to first- or second-order false belief attribution (Platek, Critton, Myers, & Gallup, 2003). Witnessing other people yawn triggers *precuneus* and *posterior cingulate* activation, both parts of the social cognition brain network, which suggests that contagious yawning may be an index of social cognition (Platek, Mohamed, & Gallup, 2005).

However, if contagious yawning is indeed a form of social cognition, it is certainly a primitive form of social cognition preserved in other species. Thus, it can be used as an index of social cognition in primate studies. Moreover, because infants between the ages of six months and three years fail to show contagious yawning effects (Millen & Anderson, 2011), contagious yawning may not reflect a purely innate social cognitive ability but, rather, be one that is acquired through social experience. Specifically, contagious yawning may have a genetic predeterminant but require specific environmental interactions or experiences (for instance learning, adaptation, or tuning) before people can express the behavior. Alternatively, other biological systems necessary for contagious yawning may simply not yet have come online or have been developed.

Contagious yawning has been observed in a variety of nonhuman species, including chimpanzees (Anderson, Myowa-Yamakoshi, & Matsuzawa, 2004), macaques (Paukner & Anderson, 2006), and dogs (Joly-Mascheroni, Senju, & Shepherd, 2008). In these studies the animal views a video of a conspecific yawning repeatedly while the experimenter tallies the number of yawns the animal makes. This tally is compared to a control condition in which the animal watches a video without a conspecific yawning. Researchers have found evidence of contagious yawning even when chimpanzees view animated versions of chimpanzees yawning (Campbell, Carter, Procter, Eisenberg, & de Waal, 2009), suggesting that the conspecifics need not be real biological agents. Other research has found that chimpanzees show contagious yawning more when watching videos of in-group members rather than out-group members (Campbell & de Wall, 2011). This result supports the social function of contagious yawning, linking it to nonconscious mimicry.

Research documents that bonobos share food with other bonobos, specifically group mates (Hare & Kwetuenda, 2010), and prefer to share food with strangers than with known group mates (Tan & Hare, 2013). Presumably, food sharing in this latter regard is motivated by ingratiation: a desire to get into the good graces of another (have the other hold a favorable opinion about oneself). In bonobo society such prosocial behavior often results in sexual interactions, and nonreproductive sex in general is used as a tool to reduce anxiety.

However, bonobo and chimpanzee societies share an important difference; bonobos are female dominant, and that perhaps accounts for greater in-group prosociality. Chimpanzees and bonobos share the same migration pattern; female chimpanzees leave their in-groups upon puberty, but unlike the non-hostile treatment received by bonobo female migrants, chimpanzee female migrants are mistreated by new in-group members. This may occur perhaps because males are dominant in chimpanzee society, resulting in secondary access to resources for females, as opposed to bonobo society where there are more abundant resources with females holding primary access. With resources limited in chimpanzee society and male dominance, female chimps do not stick together; female–female chimpanzee interactions often result in violence, and female chimpanzees often live alone.

Because bonobo females migrate to new groups upon puberty and receive rather warm receptions, bonobos are typically not averse to strangers. Even young bonobos are not stranger averse (Hohmann, 2001). Such safe migration allows new group members to continually join existing groups (Pfalzer & Ehert, 1995). Therefore, in contrast to most other primates, particularly in contrast to their very close cousin the chimpanzee (Miller, Bard, Juno, & Nadler, 1990), strangers present a potential source of reward for bonobos and are not a source of fear and trepidation.

Interestingly, it is possible to rid macaque monkeys of their stranger aversion by obliterating their amygdalae (Amaral, 2002). Such interventions have been used as animal models for autism because these monkeys show other behavior consistent with the disorder, including repetitive behaviors. In addition, some human clinical populations also lack stranger aversion, including patients who suffer from disinhibited social engagement disorder (DSED). The *Diagnostic and Statistical Manual of Mental Disorders,* 5th edition; DSM-5 (the official repository of psychiatric illness published by the American Psychiatric Association) considers DSED as a trauma- and stress-related disorder and describes it as "a pattern of behavior in which a child actively approaches and interacts with unfamiliar adults" (American Psychiatric Association, 2013). This latter example highlights the ubiquity of stranger aversion.

Chimpanzees do not engage in food sharing in the same manner as bonobos. In fact, food sharing occurs as reciprocity (de Waal, 1997). This type of tit-for-tat sharing does not constitute the same type of social cognition bestowed to the bonobos. Further evidence is provided by the environment present when chimpanzees are engaged in reciprocal food sharing: it is one of aggressive and appeasement behaviors, as well as heightened anxiety (de Waal, 1989). Clearly, conspecifics are not popular among chimpanzees, yet alone strangers, especially compared to their close genetic cousins bonobos (Hare, Melis, Woods, Hastings, & Wrangham, 2007). Nonetheless, these primates, and many others, manage to be social, live in groups, and depend on each other for day-to-day survival.

Considered together, this research suggests distinct forms of social cognition: one resulting in stranger anxiety and another resulting in stranger reward. Clearly both possibilities exist in human society, and the social context—whether walking alone in a dark alley or a waiting for a mystery date at a fancy restaurant—determines human responses to strangers. In humans, which strangers will evoke which responses remains unclear. However, bonobos and chimpanzees may provide an exaggerated depiction of both forms of social cognition. Perhaps one form may require the presence of the other to achieve balance in human societies.

Why? Gesture and Language

The analogous "why" question asks how social cognitive abilities are similar in species that did not share recent genetic similarity but did inhabit similar social contexts? Dogs live among humans—and have for millennia (Fox, 1978). Again, the question surrounds the extent to which the millennia of dog–human cohabitation have changed both human and canine social relationships. Perhaps the lower status of dogs has forced a larger change on their social relationship structure. Canines not only contagiously yawn when watching humans (Joly-Mascheroni et al., 2008), but they can infer the mental states of humans from gestures (Hare & Tomasello, 2005). Much like facial expressions or dynamic bodily movements, gestures are bodily movements that support language and social communication. These include pointing and other social hand gestures, eye gaze, bodily posture, and vocalizations. Subtle forms of behavior, facial expressions, and gestures address both the how and why of human social behavior—social communication through bodily movement.

In one classical demonstration of the remarkable social cognition skills of dogs, researchers hid food beneath one of two cups (taking special care that the food did not emit smells that could be used as cues for its location; Hare & Tomasello, 2005). They then pointed toward the cup that hid the food. While chimpanzees commonly failed such tasks, the dogs, even puppies right out of the box, were able to successfully decode the pointing behavior as a cue to the location of the food and solve this task.

A remarkable aspect of dogs is that humans helped to evolve the species' social cognitive abilities. Social cognition seems to act as a powerful environmental factor that can shape a species evolution. But are behavioral responses to human communicative gesture indications of social cognition, and if so, how did dogs acquire this difficult ability? We can begin with the latter part of the question to answer it in its entirety. Dogs evolved from campsite wolves of early human hunter-gatherers (Fox, 1978). Human campsites provided a resource such that refuse was a food source. Incidentally, trailing camp wolves also provided a service for early humans since they could warn of danger before it arrived at the campsite by howling. Over time, the more docile of the wolves interacted with the humans. Docility is a feature of infants, and abandoned wolf pups may have provide the first inclusion into human social interaction for the species. The less docile pups received less human interaction, and over time, the modern dog emerged. A remarkable experimental study in minx foxes over forty years using the same selection approach has produced a domesticated fox that behaves like domesticated wolves, dogs (Belyaev, Plyusnina, & Trut, 1985).

The remarkable occurrence of dogs, a species living in cooperation with humans throughout a large part of both species' recent evolution, suggests a few interesting hypotheses. First, spoken language is not necessary for social cognition, suggesting an answer to the second question above about the superfluous nature of language for social cognition. Perhaps social cognitions are representations that can exist without spoken language because complicated cognitive constructs are not necessary to respond to human communicated behavior or to solve other social tasks (Andrews, 2012). Infant research on emotion category learning discussed in chapter 5 further supports this hypothesis.

Second, docility is a prerequisite for social cognition acquisition. This is not to suggest that nondocile animals cannot gain social cognition ability, but I argue that such an animal must first display docile behavior in order for social cognition to have the opportunity to unfold. Stated differently, if the brain had to be in an aggressive, threatened state, then the mental resources necessary to execute social cognition might not be available. Additionally, perhaps docility allows physical contact, sparking necessary hormones such as oxytocin, vasopressin, and dopamine—hormones important for bonding and caregiving behavior present in most mammals. The subsequent interaction of these hormones within the tension created in the interspecies environment bereft of spoken language could push gesture to serve the same function as spoken language while at the same time promoting social cognition.

Nonetheless, language plays an important role in social cognition in that it serves to unify behavior around a concept. As an example, the trait "clumsy" unifies a number of behaviors around the concept of errors. Clumsy is not the only such trait, but its concept lacks the incompetence of "careless" or the tottering of "ungainly" or the harshness of "inept." It has a humorous element found in "silly" without the intentional nature of silly behaviors. Language also is useful since it indicates when a person has engaged social cognition. There are a number of different types of verbs—words to describe behavior—that differ in the level of abstractness or concreteness of the behavior (Semin & Fiedler, 1989). This continuum captures whether social cognition has occurred because the most abstract verbs, called *mental state verbs,* suggest the occurrence of a mental state inference.

Modern Environmental Pressures

The evolutionary anthropologists tell us that despite the remarkable abilities of nonhuman primates, humans are still a cut above the rest in terms of social cognition (Hare, 2011). Human babies show social cognitive skill much earlier than any other species, presumably to allow integration not only in a social group but into a culture as well. Human social life is much more complex than animal social life, and this additional complexity may have been a driving force evolving human social cognitive abilities, pushing us to decipher mental states relatively early in life (Humphrey, 1976). These scientists also tell us that these social cognitive abilities are built on the abilities of an early primate ancestor, suggesting that social context has always been important since before human beings took their modern forms. This suggests that social cognition is an evolutionarily old ability that probably is the scaffolding on which modern intelligence is built. I discuss this in more detail in chapter 7; for now, let us revisit the idea of the physical environment to paint a more comprehensive picture of the term "environment."

Modern technology has made social communication possible across the globe. As a result, physical contact during social relationships is not necessary to engage social communication; bodily movements and spoken language are still possible to communicate, e.g., via Skype—but with someone else on the other side of the world. This has minimized the impact of the physical environment on social interactions and suggests that the technological advance itself is geared toward enhancing the experience of this important feature of human social life. Therefore, we have arrived at the conclusion that social communication (language) provides the mental construct tension that enables an important aspect of flexible social cognition. I admit that it is unclear whether other animals that lack language also display flexible social cognition. Nonetheless, I continue to consider this concept throughout the book and to revisit reward, fear, and learning in the next chapter. Now, I also turn the focus to other forms of social behavior to continue our explanation of the *how* and *why* questions surrounding social cognition.

References

Amaral, D. G. (2002). The primate amygdala and the neurobiology of social behavior: Implications for understanding social anxiety. *Biological Psychiatry*, 51(1), 11–17.

American Psychiatric Association. (2013). *Diagnostic and statistical manual of mental disorders* (5th ed.). Washington, DC: APA.

Anderson, J. R., Myowa-Yamakoshi, M., & Matsuzawa, T. (2004). Contagious yawning in chimpanzees. *Proceedings. Biological Sciences*, 271(7), S468–S470.

Andrews, K. (2012). *Do apes read minds? Toward a new folk psychology*. Cambridge, MA: MIT Press.

Belyaev, D. K., Plyusnina, I. Z., & Trut, L. N. (1985). Domestication in the silver fox (*Vulpes fulvus*, Desm): Changes in physiological boundaries of the sensitive period of primary socialization. *Applied Animal Behaviour Science*, 13(4), 359–370.

Campbell, M. W., Carter, J. D., Proctor, D., Eisenberg, M. L., & de Waal, F. B. (2009). Computer animations stimulate contagious yawning in chimpanzees. *Proceedings of the Royal Society B: Biological Sciences*, 276(1676), 4255–4259.

Campbell, M. W., & de Waal, F. B. (2011). Ingroup-outgroup bias in contagious yawning by chimpanzees supports link to empathy. *PLoS One*, 6(4), e18283.

Carter, C. S. (1998). Neuroendocrine perspectives on social attachment and love. *Psychoneuroendocrinology*, 23(8), 779–818.

Carver, C. S., & Harmon-Jones, E. (2009). Anger is an approach-related affect. Evidence and implications. *Psychological Bulletin*, 135, 183–204.

Chang, R. S. (2013). Human attachment vocalizations and the expanding notion of nurture. In M. L. Fisher, J. R. Garcia, & R. S. Chang (Eds.), *Evolution's empress: Darwinian perspectives on the nature of women* (pp. 168–187). New York: Oxford University Press.

Churchland, P. S., & Winkielman, P. (2012). Modulating social behavior with oxytocin: How does it work? What does it mean? *Hormones and Behavior*, 61(3), 392–399.

Darwin, C. (1859). *The origin of species by means of natural selection, or, the preservation of favoured races in the struggle for life* (Vol. 2). London: International Science Library.

de Waal, F. M. (1997). The chimpanzee's service economy: Food for grooming. *Evolution and Human Behavior*, 18(6), 375–386. doi:10.1016/S1090–5138(97)00085–8.

de Waal, F. W. (1989). Food sharing and reciprocal obligations among chimpanzees. *Journal of Human Evolution*, 18(5), 433–459.

Dunsmoor, J. E., Murty, V. P., Davachi, L., & Phelps, E. A. (2015). Emotional learning selectively and retroactively strengthens memories for related events. *Nature*, 520(7547), 345–348.

Ferneyhough, E., Kim, M. K., Phelps, E. A., & Carrasco, M. (2013). Anxiety modulates the effects of emotion and attention on early vision. *Cognition and Emotion*, 27(1), 166–176.

Fox, M. W. (1978). *The dog: Its domestication and behavior.* New York: Garland Publishing.

Gingrich, B., Liu, Y., Cascio, C., Wang, Z., & Insel, T. R. (2000). Dopamine D2 receptors in the nucleus accumbens are important for social attachment in female prairie voles (*Microtus ochrogaster*). *Behavioral Neuroscience*, 114(1), 173.

Gray, P. B., & Anderson, K. G. (2010). *Fatherhood: Evolution and human paternal behavior*. Cambridge, MA: Harvard University Press.

Haidt, J. (2001). The emotional dog and its rational tail: A social intuitionist approach to moral judgment. *Psychological Review*, 108(4), 814.

Haidt, J., Rozin, P., McCauley, C., & Imada, S. (1997). Body, psyche, and culture: The relationship between disgust and morality. *Psychology and Developing Societies*, 9(1), 107–131.

Hare, B. (2011). From hominoid to hominid mind: What changed and why? *Annual Review of Anthropology*, 40, 293–309.

Hare, B., & Kwetuenda, S. (2010). Bonobos voluntarily share their own food with others. *Current Biology*, 20(5), R230–R231.

Hare, B., Melis, A. P., Woods, V., Hastings, S., & Wrangham, R. (2007). Tolerance allows bonobos to outperform chimpanzees on a cooperative task. *Current Biology*, 17(7), 619–623.

Hare, B., & Tomasello, M. (2005). Human-like social skills in dogs? *Trends in Cognitive Sciences*, 9(9), 439–444.

Hartley, C. A., & Phelps, E. A. (2012). Anxiety and decision-making. *Biological psychiatry*, 72(2), 113–118.

Heider, F. (1946). Attitudes and cognitive organization. *Journal of Psychology*, 21, 107–112.

Hohmann, G. (2001). Association and social interaction between strangers and residents in Bonobos (*Pan paniscus*). *Primates*, 42(1), 91–99. doi:10.1007/ BF02640692.

Humphrey, N. (1976). The social function of intellect. In P. P. G. Bateson & R. A. Hinde (Eds.), *Growing points in ethology* (pp. 303–317). Cambridge, UK: Cambridge University Press.

Joly-Mascheroni, R. M., Senju, A., & Shepherd, A. J. (2008). Dogs catch human yawns. *Biology Letters*, 4(5), 446–448.

Kling, A. (1975). Testosterone and aggressive behavior in man and non-human primates. *Hormonal Correlates of Behavior*, 1, 305–326.

Lindquist, K. A., Wager, T. D., Kober, H., Bliss-Moreau, E., & Barrett, L. F. (2012). The brain basis of emotion: A meta-analytic review. *Behavioral and Brain Sciences*, 35(03), 121–143.

Mashour, G. A., & Alkire, M. T. (2013). Evolution of consciousness: Phylogeny, ontogeny, and emergence from general anesthesia. *Proceedings of the National Academy of Science, USA,* 110(Suppl 2), 10357–10364. doi: 10.1073/pnas.1301188110.

Millen, A., & Anderson, J. R. (2011). Neither infants nor toddlers catch yawns from their mothers. *Biology Letters*, 7(3), 440–442.

Miller, L. C., Bard, K. A., Juno, C. J., & Nadler, R. D. (1990). Behavioral responsiveness to strangers in young chimpanzees (*Pan troglodytes*). *Folia Primatologica*, 55(3–4), 142–155. doi:10.1159/000156510.

Paukner, A., & Anderson, J. R. (2006). Video-induced yawning in stumptail macaques (*Macaca arctoides*). *Biology Letters* , 2, 36–38. doi:10.1098/rsbl.2005.0411.

Pfalzer, S., & Ehert, G. (1995). Social integration of a bonobo mother and her dependent daughter into an unfamiliar group. *Primates*, 36(3), 349–360. doi:10.1007/BF02382858.

Platek, S. M., Critton, S. R., Myers, T. E., & Gallup, G. G. (2003). Contagious yawning: The role of self-awareness and mental state attribution. *Brain Research. Cognitive Brain Research*, 17(2), 223–227.

Platek, S. M., Mohamed, F. B., & Gallup, G. G. (2005). Contagious yawning and the brain. *Cognitive Brain Research*, 23(2), 448–452.

Pochedly, J. T., Widen, S. C., & Russell, J. A. (2012). What emotion does the "facial expression of disgust" express? *Emotion*, 12(6), 1315–1319. doi:10.1037/a0027998.

Rosati, A. G., Santos, L. R., & Hare, B. (2012). *Primate social cognition: Thirty years after Premack and Woodruff*. In: M. L. Platt & A. S. Ghazanfar (Eds.), *Primate neuroethology* (pp. 117–143). London, UK: Oxford University Press.

Ross, H. E., Cole, C. D., Smith, Y., Neumann, I. D., Landgraf, R., Murphy, A. Z., et al. (2009). Characterization of the oxytocin system regulating affiliative behavior in female prairie voles. *Neuroscience*, 162(4), 892–903.

Ross, H. E., & Young, L. J (2009). Oxytocin and the neural mechanisms regulating social cognition and affiliative behavior. *Frontiers in Neuroendocrinology*, 30(4), 534–547.

Semin, G. R., & Fiedler, K. (1989). Relocating attributional phenomena within a language-cognition interface: The case of actors' and observers' perspectives. *European Journal of Social Psychology*, 19(6), 491–508.

Tan, J., & Hare, B. (2013). Bonobos share with strangers. *PLoS One*, 8(1), e51922.

Thomsen, D. K. (2006). The association between rumination and negative affect: A review. *Cognition and Emotion*, 20(8), 1216–1235.

Walker, R. S., Hill, K. R., Flinn, M. V., & Ellsworth, R. M. (2011). Evolutionary history of hunter-gatherer marriage practices. *PLoS One*, 6(4), e19066.

Wang, Z., Yu, G., Cascio, C., Liu, Y., Gingrich, B., & Insel, T. R. (1999). Dopamine D2 receptor-mediated regulation of partner preferences in female prairie voles (*Microtus ochrogaster*): A mechanism for pair bonding? *Behavioral Neuroscience*, 113(3), 602.

Warneken, F., & Tomasello, M. (2009). Varieties of altruism in children and chimpanzees. *Trends in Cognitive Sciences*, 13(9), 397–402.

Whalen, P. J., & Phelps, E. A. (2009). *The human amygdala*. Guilford Press.

Widen, S. C., & Russell, J. A. (2010). The "disgust face" conveys anger to children. *Emotion*, 10(4), 455–466. doi:10.1037/a0019151.

Widen, S. C., & Russell, J. A. (2013). Children's recognition of disgust in others. *Psychological Bulletin*, 139(2), 271–299. doi:10.1037/a0031640.

Williams, J. R., Insel, T. R., Harbaugh, C. R., & Carter, C. S. (1994). Oxytocin administered centrally facilitates formation of a partner preference in female prairie voles (*Microtus ochrogaster*). *Journal of Neuroendocrinology*, 6(3), 247–250.

Yoon, J. M., & Tennie, C. (2010). Contagious yawning: A reflection of empathy, mimicry, or contagion? *Animal Behaviour*, 79(5), e1–e3.

Young, L. J., Lim, M. M., Gingrich, B., & Insel, T. R. (2001). Cellular mechanisms of social attachment. *Hormones and Behavior*, 40(2), 133–138.

3 Prediction

Imagine a day without knowing the basic properties of physical objects; simply getting out of bed might prove difficult. On waking, you might wish to roll out from under the covers, but you would not know that the floor was solid and capable of supporting your weight. Given this state of uncertainty, you might think that you could even float away when you pulled the covers off without knowing that gravity (not the covers) kept you comfortably beneath the sheets. Knowing that the floor is solid and capable of supporting your weight, knowing that you will not float up to the ceiling once the covers were removed because gravity keeps you firmly grounded allows you to execute the simple act of getting out of bed. All this information about physical objects—the floor and your body—allows you to expect that when you pull the covers off, you would remain in bed, or when you put your feet down, the floor would support your weight (although you might also expect to feel a shiver if you did not step directly onto your slippers).

Prediction is a basic survival instinct. Knowing the basic properties of physical objects allows you to *predict* the result of your interaction with them, facilitating your navigation of the physical world. The same is true for your interactions with people and other social agents. Every action that an organism engages carries consequences, and the ability to predict those consequences increases the probability of the organism achieving its primary goal: survival.

Predicting the behavior of agents relies on information: both of the actions that the agent is capable of executing and the properties of other physical objects and social agents in the environment, as well as a sense of timing; expectations about when specific things should occur in relation to other things. Knowledge of the properties of physical objects and social agents is acquired via a *Bayesian* process requiring the gathering of statistical information that itself depends on prediction. However, unlike acquiring knowledge of physical objects, acquiring knowledge of social agents drives social cognition.

An organism acquires knowledge about the external world during its lifetime through a Bayesian process (see, for example, Mangel, 1990). This requires integrating statistical information (means and variance) to form expectations about the likely occurrence of events. More specifically, people estimate a posterior given the likelihood, priors, and evidence. The evidence comes from observed behavior, while the priors and likelihood are expectations based on previous behavior, social knowledge, and best guesses. These expectations can be considered base rates, and they describe the state of the world at any given moment from the organism's point of view. However, decades of research have long demonstrated that people sometimes ignore base rates when making decisions because such information anchors their choices (Kahneman & Tversky, 1983). For example, if people are informed that Linda is a bank teller who also engages in behavior consistent with being feminist, and are then asked to estimate the likelihood that Linda is a feminist bank teller versus simply a bank teller, most people would respond that Linda is the former, although the latter is statistically more likely (there are simply more bank tellers than there are feminist bank tellers).

Other phenomena call into question whether people are always Bayesian in their prediction or, rather, whether people use other features of the statistical distribution. For instance, if asked to report their birthdate then to estimate the number of people who died from Ebola this year, most people would be influenced in their Ebola estimate by their birthdate because this number would serve as an anchor from which they failed to sufficiently adjust. This suggests that the anchor need not be something related to the prediction, and moreover, the predictive mechanism probably functions unconsciously since it can be influenced by irrelevant information.

It should be noted that since social cognition is cognitively costly, we rely on a heuristic to determine when it should be engaged. That heuristic can be a social norm present in a context. If we notice someone deviating from normative behavior, that deviation may be sufficient to present a tug on our metaphorical fishing line, allowing us to abandon the norm as the predictor and consider their unique mental states. But for most of our interactions in social situations with multiple people, each with multiple mental states to consider, a heuristic (such as social norms) may be best, particularly initially. This heuristic provides a prediction akin to predicting the behavior of objects: it is an effortless prediction that occurs nonconsciously, allowing navigation of the physical environment (which often is populated by other people). For instance, at a restaurant, I may, over time, begin imagining the mental states of the other customers sitting nearby. But initially, on entering and sitting, ordering, and chatting with my companion, I may not consider their minds.

Social Prediction

All people share some features, some people share most features, and each individual has unique features. Given the complexity of humanity and human mental states, knowing when to instantiate social cognition to facilitate a specific prediction about a person in a unique situation may require a bit of "magic": statistical magic. The idea that people can use statistics to create a reality consistent with their interests, policies, beliefs, desires, and past behaviors has existed for some time (for one such discussion, see Holmes, 1990). Again, the motive for consistency in humans is relevant.

Since human beings use statistical inferences in their Bayesian approach to social cognition, then it is possible they are also capable of abusing statistics in this manner. Social prediction is an example of such abuse. Features that most people share can be considered base rates. These tacit predictions are usually nonconscious and allow us to navigate our social environment in a simple, noncomplex manner. However, base rates can also be gleaned from the social context; norms are rules of behavior that provide statistical information about how a person, on average, should behave in a given situation (Smith, Hilton, Kim, & Garonzik, 1992). Norms and other social conventions are culturally determined, and they require social learning. More importantly, norms may be more reliable predictors of behavior since they describe behavior on average. And they do not require social cognition.

The use of base rates even extends to human behavior in the form of rules about behavior. If someone is kicked, that person will experience pain and possibly vocalize that experience. If someone meets a close relation after some time apart, she or he will hug that individual. Therefore, the probability that someone will do something is close to 100% (or 1 in our earlier mathematical formulation) if that person is in the relevant social context. This latter statement suggests that this type of base rate application creates expectations about how social contexts drive people's behavior. Although it is well documented that people ignore the social context or situation when *explaining* behavior (for an interesting summary, see Darley & Cooper, 1998), the situation is prevalent during *prediction* even though people may be unaware of its influence. The reason people can share libraries and public transport and can attend large concerts and religious events is that each individual in that environment is tacitly aware of how each other individual will behave in that same environment. Violations of these expectations can lead to ostracism, imprisonment, and general social exclusion. These predictions are usually about things external to the person.

Social psychology has argued for decades that trait inferences, a form of social cognition, are used to predict behavior. However, this debate, first sparked when modest correlations of on average +0.30 were found between traits and behavior, has often ignored a distinction between prediction and explanation (see Bem & Allen, 1974, for an early review). Instead, prediction and explanation have been considered two sides of the same coin or process. Thus, when asking people to use their explanatory powers to report what trait they inferred from behavior, researchers assume these same traits are used as predictive tools to guide these behaviors. This is indeed problematic if one holds the view that prediction and explanation depend on separate neural and cognitive mechanisms. This is not to say that there is no overlap or communication between predictive and explanatory mechanisms, but for the purposes of this discussion, I consider them as separable mechanisms.

To test whether traits guide future behaviors without the occurrence of explanation, we devised a study in which we tested the complete process from obtaining behavioral information about a person through to predicting what that person would do in a future, seemingly related situation (Harris, Lee, Thompson, & Kranton, 2015). We used consequential, actual decisions made in the context of an economic paradigm not only to motivate behavior but to allow assessment of normative behavior: a base rate of the social context. We also asked participants to report their perceived norms within the context. Additionally, we collected from a separate sample trait ratings of the behavioral information, allowing us to test whether traits indeed predicted behavior if they were not the result of self-report, triggering explanatory mechanisms. Although we found that norms did predict behavior, along with participants' estimates of the ideal norms in the social context, traits did not predict behavior. Thus, traits did not generalize. Admittedly, other traits not tested in this study may indeed generalize, but the core person perception traits tested in this study, warmth and competence, failed to do so.

In support, striking examples of base rate use come from social roles (Zukier & Pepitone, 1984), stereotypes (Davidson & Hirtle, 1990), and personality traits (Ginosar & Trope, 1980). People expect priests to preach, mothers to nurture, teachers to teach, and grumpy people to have fewer friends. Therefore, the expectation that a particular priest should preach is available for use. This latter, more restricted use of base rates highlights the fact that some people share most features. Whether tall or short, dark-, light-, yellow-, red-, or brown-skinned, a teacher will still teach. Another view of this use of base rates is stereotyping. Unlike predictions about all people (which may be considered people stereotypes) social group stereotypes reside within a subset of people defined by the perceived social group, role, or culture of that individual.

Group boundaries serve to instantiate cognitive processing based on base rates. People usually make the same prediction for the same situation each time, but these predictions can vary based on perceived fit to social role or membership within the stereotyped category. For instance, a bad teacher may not be expected to teach all the time and probably often escapes to the teachers' lounge for the most of her class to surf the Internet (just as her students commonly do in class). Or a Catholic who is not particularly religious may not be expected to sing the hymns along with everyone else at Sunday mass because she does not remember the words and is not particularly motivated to pick up the hymnal, look up the song, and join in. Therefore, there is room for variability from 100% (or 1) based on these appraisals. These predictions are about things internal to the person.

But how could situation-specific base rates or *norms* predict behavior, given that they represent information about the social context, not individual difference information about the specific person? Since the social context guides behavior, it makes logical sense that base-rate information about that context is informative. If we know what, on average, any person is likely to do in a given situation, we can feel fairly confident in predicting the behavior of a person in that situation. This individual difference is captured by simply considering the extent to which the person deviates from the average person's behavior in a related context. The deviation can be above or below the mean to varying degrees, capturing the full extent of individual difference in the manner of a trait inference. Such inferences allow predictions about the interaction of the situation and the individual. These are the unique predictions to individuals based on who they are and the social context in which their behavior takes place. Therefore, traits specifically and social cognitions more broadly are not necessary to predict behavior, and may satisfy only social motives related to social interaction and communication (language), the domain of explanation.

A final consideration on social prediction comes from economists—and not just any economists, but those who have integrated social psychology theory into their economic models. The Akerlof and Kranton (2000) model integrates social identity theory in social psychology (Tajfel & Turner, 1979), which describes the process by which people belong to multiple social identities tied to multiple social groups. However, instead of simply reconstituting economic theory, Akerloff and Kranton integrate social identity as an additional argument in a classical utility model. Akerlof and Kranton argue that a person's behavior in a situation, the behavior of others in the situation, and the person's social identity, all of which are governed by parameters of the social relationships (conventions or norms), and placement of all these people in the relevant social categories, are factored into the utility function. They further argue that there are two types of factors that drive placement in a social category: *exogenous and endogenous factors*. Those that are endogenous are specific to the social context, whereas those that are exogenous are specific to anything but the context. Interestingly, this economic approach also places a premium on the social context as well as on norms that govern said context.

Prediction Violation

As hinted above, predictions often go awry. This feature of predictions is essential for learning—people update their expectations based on incorrect predictions. This *prediction error* is a concept extensively discussed in the decision-making and learning literatures (Schultz, 1998). Researchers have long identified *dopamine neurons* in the *striatum* in nonhumans that track prediction error: these cells fire when they encounter prediction error but do not fire when the expectation is met (Schultz, Dayan, & Montague, 1997). The striatum is a collection of structures in the subcortex and includes the caudate, putamen, and nucleus accumbens. Decades of research in humans have confirmed the role of dopamine and the striatum in generating prediction error signals, suggesting that this brain system plays a vital role in guiding behavior.

Do these prediction error brain signals also occur in social contexts? Specifically, does the striatum also underlie violations of social expectations? Previous research had associated other brain regions like the anterior cingulate cortex (ACC) in expectation violations (see Potts, Martin, Kamp, & Donchin, 2011, for comparison of electrophysiological error signals localized to the ACC), but perhaps the striatum may still have a role to play in social prediction. In an experimental test of this hypothesis, participants were instructed to make dispositional inferences about people before violating expectations (Harris & Fiske, 2010). For instance, if I told you that John reads to sick children at the hospital, he does so when hardly anyone else does, he is willing to read to any other patient at the hospital, and he has always read to sick children in the hospital, you might begin conjuring an image of a kind, empathic, compassionate person. These traits are associated with people in specific social roles such as nurses and people who belong to specific social groups such as the elderly. Therefore, it is highly likely that, on hearing a description of John's behavior, you brought to mind a person consistent with these social roles or stereotypes. So did our participants. However, following the description of John's behavior, we showed participants a picture of John that violated their expectation (John was depicted as shooting up heroin, fully looking the part of a drug addict). Since we do not associate the traits mentioned above with drug addicts, this should create an expectation violation and trigger appropriate brain processing.

We found that even in the face of the expectation violation, participants still made the trait attribution to John, evidenced by self-report ratings and engagement of the social cognition brain network. However, the unexpected image of John also engaged both the caudate and putamen, brain structures that are part of the striatum and implicated in learning. Moreover, this effect held even when the trait domain switched from morality–warmth to competence, suggesting that the brain activation reflects a general violation signal, not one specific to kind behavior. These data provided the first evidence that social prediction error relies on the same system as prediction error in other, decision-making or economic contexts. Moreover, these data implicate the general reward system in social cognition.

Social Reward

But is the reward system really engaged in social contexts as it is in economic contexts? Certainly social stimuli are rewarding, and the presence of the person positivity bias—we tend to perceive people as slightly positive—suggests that other people present an inherent source of reward (Taylor & Brown, 1988). However, this literature also suggests that social perception presents additional elements (Fiske & Taylor, 2013). For instance, people as stimuli are unique because another person will form an impression of you while you are forming an impression of her. Since we care about what other people think of us, we are constantly attempting to manage the impression the other person forms while we evaluate them ourselves. That should require additional processing beyond simple evaluation. Secondly, other people are intentional agents, actively influencing the environment. Given that they are not simply passive objects, this intentional feature of human behavior also has to be accounted for and may influence reward processing. Finally, it is difficult to verify the accuracy of social cognition because people can engage in deception (see Sugiyama, Tooby, & Cosmides, 2002; Trivers, 1971, for a similar account): they can behave in a manner not consistent with their beliefs, desires, and motives. This may also impact reward processing. Related to this point is the fact that people possess private mental states that are known only to them. For all of these reasons, social reward processing should be fundamentally different from general reward processing.

We decided to test whether social rewards engaged the same brain regions as other rewards. We attempted this over two very different paradigms. In the first paradigm we simply showed participants images of positive and negative objects and traditionally dehumanized or not traditionally dehumanized people (Harris, McClure, Van den Bos, Cohen, & Fiske, 2007). Their task was simply to categorize the images as positive or negative and human or not. The second study had participants perform a time estimation task in which they guessed how long a certain length of time (9.5 seconds) took to elapse (Van den Bos, McClure, Harris, Fiske, & Cohen, 2007). They received feedback on their performance either from the experimenter or from the computer program running the experiment. This feedback was either positive or negative in the form of juice or quinine (bitter liquid used in tonic water) respectively squirted into their mouths. Across both studies, regions of medial prefrontal cortex (MPFC) predicted valence differences between the stimuli, more active for positive than negative. However, in some regions of MPFC, including regions superior, inferior, and anterior to the horn of the cingulate, as well as the subgenual cingulate, we found a valence × human interaction whereby positive or negative humans (whether categorizing or receiving feedback from) drove activation much more strongly than nonhumans. This hints at the function in the M_t equation, suggesting that social reward could be additive because the brain region responded to rewards but increased its response to social rewards. Therefore, social rewards may rely on similar reward mechanisms that govern nonsocial rewards, although there is an additive affect when mental states are considered. Stated differently, the additional processing necessary to infer mental states may influence the reward mechanism, causing more or less reliance on it to govern the decision. Therefore, it is logically possible that a social reward may not drive the reward system but simply be represented abstractly along with the mental state inference.

Social Punishment

Apart from rewards, punishments also factor into predictions. This "carrot and stick" approach ensures that people are well motivated to make accurate predictions to gain rewards but avoid punishments. However, neuroscience research has revealed that punishment may not be the other side of the reward coin: instead, punishment depends on a wider brain system than reward, although there is some overlap (Seymour, Singer, & Dolan, 2007). In addition to the areas implicated in reward such as the striatum and prefrontal cortex (medial and lateral orbitofrontal cortex [MOFC, LOFC], and MPFC), punishment involves the insula, the amygdala, and periaqueductal gray (Seymour et al., 2007). These latter, subcortical brain regions are also involved in affective processing, particularly of negative affect. The amygdala plays a key role in fear learning and social biases. The insula and putamen are necessary to experience disgust. These additional influences highlight the role of negative affect in punishment.

Punishment often relies on anger, and it can be motivated by revenge (Seip, Rotteveel, van Dillen, & van Dijk, 2014). This occurs both in second-party contexts where the punisher has been wronged and in third-party contexts where people punish on behalf of a victim. Interestingly, third-party punishment relies on the social cognition brain network, even when it is at a cost to the punisher (de Quervain et al., 2004). This suggests that such punishment is inherently social and requires thoughts about not just the punished but the victim and perhaps has implications for the social group or society.

Interestingly, considering reward and punishment in concert raises questions about other types of rewards and punishment. For instance, is the absence of punishment (relief) rewarding, and is the absence of a reward (disappointment) punishing? Fewer neuroscience studies have addressed these questions, but there is some research to discuss. For instance, researchers have demonstrated that fear of losing money operates like a conditioned stimulus in a fear learning paradigm driving physiological responses (Delgado, Labouliere, & Phelps, 2006). This suggests that expectations about the context (expecting to gain or lose) may determine when the reward system or the more extended punishment system becomes engaged.

Final Thoughts on Prediction

This chapter has argued that social cognition is not necessary for predicting people's behavior. This does not mean that it is not sufficient for this task, simply that people use other heuristics such as norms to predict other people's behavior most of the time. Seemingly halo effects (Thorndike, 1920), spontaneous trait inferences (Newman & Uleman, 1990), and other similar phenomena document instances where people do use trait inferences to predict. However, this seems to be the case when explanatory mechanisms are likely to be co-active, suggesting that traits may serve as predictive heuristics only when explanation is also necessary. Most of our implicit predictions occur without any need for explanation, including implicit predictions about people based on stereotypes, social roles, social conventions, norms, mores, and cultural prescriptions.

Finally, it is important to note that social cognition is not necessary for predicting people's behavior, although its use will not always lead to incorrect predictions. Stated differently, social cognition may be sufficient for accurate predictions but is not necessary. However, taking prediction away from social cognition does not devalue its importance in social interactions. In fact, given the importance of consistency and sense making to human functioning, social cognition remains a prominent ingredient in the human experience. Perhaps this distinction will free social cognition researchers to explore the phenomenon as a more conceptual occurrence that drives behavior. Next, I discuss explanation, the domain of social cognition.

References

Akerlof, G., & Kranton, R. (2000). Economics and identity. *Quarterly Journal of Economics*, CVX(3), 715–753.

Bem, D. J., & Allen, A. (1974). On predicting some of the people some of the time: The search for cross-situational consistencies in behaviours. *Psychological Review*, 81, 506–520.

Darley, J. M., & Cooper, J. (1998). *Attribution and social interaction: The legacy of Edward E. Jones.* Washington, DC: American Psychological Association. 10.1037/10286-000.

Davidson, D., & Hirtle, S. C. (1990). Effects of nondiscrepant and discrepant information on the use of base rates. *American Journal of Psychology*, 103(3), 343–357. doi:10.2307/1423214.

Delgado, M. R., Labouliere, C. D., & Phelps, E. A. (2006). Fear of losing money? Aversive conditioning with secondary reinforcers. *Social Cognitive and Affective Neuroscience*, 1(3), 250–259.

De Quervain, D. J. F., Fischbacher, U., Treyer, V., Schellhammer, M., Schnyder, U., Buck, A., et al. (2004). The neural basis of altruistic punishment. *Science*, 305(5688), 1254–1258.

Fiske, S. T., & Taylor, S. E. (2013). *Social cognition: From brains to culture* . Thousand Oaks, CA: *Sage*.

Ginosar, Z., & Trope, Y. (1980). The effects of base rates and individuating information on judgments about another person. *Journal of Experimental Social Psychology*, 16(3), 228–242. doi:10.1016/0022-1031(80)90066-9.

Harris, L. T., & Fiske, S. T. (2010). Neural regions that underlie reinforcement learning are also active for social expectancy violations. *Social Neuroscience*, 5, 76–91.

Harris, L. T., Lee, V. K., Thompson, E., & Kranton, R. (2015). Norms, not traits, predict behaviour. *Journal of Behavioral Decision Making*. Online in advance of publication.

Harris, L. T., McClure, S., Van den Bos, W., Cohen, J. D., & Fiske, S. T. (2007). Regions of MPFC differentially tuned to social and nonsocial affective stimuli. *Cognitive and Behavioral Neuroscience*, 7, 309–316.

Holmes, C. B. (1990). *The honest truth about lying with statistics*. Springfield, IL: Charles C. Thomas.

Kahneman, D., & Tversky, A. (1983). Can irrationality be intelligently discussed? *Behavioral and Brain Sciences*, 6(3), 509–510. doi:10.1017/S0140525X00017246.

Mangel, M. (1990). Dynamic information in uncertain and changing worlds. *Journal of Theoretical Biology*, 146(3), 317–332.

Newman, L. S., & Uleman, J. S. (1990). Assimilation and contrast effects in spontaneous trait inference. *Personality and Social Psychology Bulletin*, 16(2), 224–240. doi:10.1177/0146167290162004.

Potts, G. F., Martin, L. E., Kamp, S. M., & Donchin, E. (2011). Neural response to action and reward prediction errors: Comparing the error-related negativity to behavioral errors and the feedback-related negativity to reward prediction violations. *Psychophysiology*, 48(2), 218–228.

Schultz, W. (1998). Predictive reward signal of dopamine neurons. *Journal of Neurophysiology*, 80, 1–27.

Schultz, W., Dayan, P., & Montague, P. R. (1997). A neural substrate of prediction and reward. *Science*, 275, 1593–1599.

Seip, E. C., Rotteveel, M., van Dillen, L. F., & van Dijk, W. W. (2014). Schadenfreude and the desire for vengeance. In W. W. van Dijk & J. W. Ouwerkerk (Eds.), *Schadenfreude: Understanding pleasure at the misfortune of others* (pp. 227–241). New York: Cambridge University Press. doi:10.1017/CBO9781139084246.019.

Seymour, B., Singer, T., & Dolan, R. (2007). The neurobiology of punishment. *Nature Reviews Neuroscience*, 8, 300–311.

Smith, R. H., Hilton, D. J., Kim, S. H., & Garonzik, R. (1992). Knowledge-based causal inference: Norms and the usefulness of distinctiveness. *British Journal of Social Psychology,* 31(3), 239–248. doi:10.1111/j.2044-8309.1992.tb00968.x.

Sugiyama, L. S., Tooby, J., & Cosmides, L. (2002). Cross-cultural evidence of cognitive adaptations for social exchange among the Shiwiar of Ecuadorian Amazonia. *Proceedings of the National Academy of Sciences of the United States of America,* 99(17), 11537–11542.

Tajfel, H., & Turner, J. C. (1979). An integrative theory of intergroup conflict. *Social Psychology of Intergroup Relations,* 33(47), 74.

Taylor, S. E., & Brown, J. D. (1988). Illusion and well-being. A social psychological perspective on mental health. *Psychological Bulletin,* 103(2), 193–210. doi:10.1037/0033-2909.103.2.193.

Thorndike, E. L. (1920). A constant error in psychological ratings. *Journal of Applied Psychology,* 4(1), 25–29.

Trivers, R. L. (1971). The evolution of reciprocal altruism. *Quarterly Review of Biology,* 46(1), 35–57.

Van den Bos, W., McClure, S., Harris, L. T., Fiske, S. T., & Cohen, J. D. (2007). Dissociating affective evaluation and social cognitive processes in ventral medial prefrontal cortex. *Cognitive and Behavioral Neuroscience,* 7, 337–346.

Zukier, H., & Pepitone, A. (1984). Social roles and strategies in prediction: Some determinants of the use of base-rate information. *Journal of Personality and Social Psychology,* 47(2), 349–360. doi:10.1037/0022-3514.47.2.349.

4 Explanation

The bigger triangle was in control or trying to take control of the smaller triangle and the circle; the rectangular shaped place was similar to like a room with a closed door that, um, if you went in there you were safe until that triangle came in. The small triangle and the circle were trying to escape from the large triangle, and when they did, the large triangle became very furious and destroyed things.
—Normal control subject

OK, so, a rectangle, two triangles, and a small circle. Let's see, the triangle and the circle went inside the rectangle, and then the other triangle went in, and then the triangle and the circle went out and took off, left one triangle there. And then the two (pause) parts of the rectangle made like a [*sic*] upside-down V, and that was it.
—Patient with bilateral amygdala legion as described by Heberlein and Adolphs (2004)

People often spend parts of the day wondering why; they seek explanations for the things that occur. Some forms of meaning making in primitive and modern cultures attribute causation for uncontrollable events to a supernatural being or beings (for example see Waytz, Morewedge, Epley, Monteleone, Gao, & Cacioppo, 2010). This anthropomorphized agent can usually be appeased, ingratiated, and related to like a human being, allowing human norms, rules, morals, and social conventions to be applied in an attempt to explain why things occur and influence outcomes. Rain dances, offerings of money and food to deities, and even sacrificing human and nonhuman life have all occurred throughout human history to satisfy the need for explanation. This makes explanation perhaps the most distinguishing feature of human experience.

Explanation may be a byproduct of consciousness. Philosophers have often self-reflected on what it is like to have a conscious self. This form of meta-cognition is prized as a uniquely human ability; it allows us to self-reflect or think about an abstract entity some philosophers do not even believe exists—the self. In fact, having a self has been deemed a curse by social psychologists (Leary, 2004), not the only social cognition construct to be thought of as burdensome in some way (for historical account, see North & Fiske, 2012). Perhaps one burden derived from having a self and a wider consciousness is the need to explain the events around oneself. Social cognition provides a handy *explanation*, regardless of accuracy, that may satisfy such a need.

The explanation of the tension (M_t) arising between unexplained occurrences, particularly negative events and behaviors, and a need to satisfy a concept of the self because of consciousness, is straightforward:

$$M_t = f(a, s),$$

where a is the person's behavior, and s is the social context. This can be considered a *why tension* (Why_t). For instance, suppose someone fell off the deck of a boat into a lake while the boat was anchored at a dock in calm waters. An attribution based on a trait of clumsy can be made almost immediately. This trait inference satisfies the Why_t since the agent gains a positive value in the equation. Similarly, in another social context—such as all-hands-on-deck trying to keep the boat afloat in a storm with rough, high waves—the attribution instead may go to the context as dangerous. This situational attribution also satisfies Why_t, making the s value in the equation non-0. It is also less likely that a clumsy attribution is made to the person, though not impossible (if all the people on the boat are experienced seafarers who have been through similar rough waters many times before, but the person who falls off does so frequently while no one else does, she may still be considered as clumsy). Perhaps empathic concern, sympathy, and other human emotions may be directed at the person falling overboard in the storm, but ridicule to the person in calm waters (or the clumsy person in the storm). In either case, a trait inference satisfies the need for explanation because it suggests the cause of the occurrence (the person falling off the boat) is either the person or the situation.

Having satisfied Why_t, human beings are allowed to function in a most uniquely human way; this ultrasocial animal is able to change its own behavior based on the solution to Why_t. For instance, suppose we were nervous about being on a boat. Hearing about a negative occurrence, such as a person falling off a boat, may further solidify conscious explanations or rationales for fears about being on boats. However, if we made the attribution for the occurrence to the person as clumsy, those fears may be somewhat subsided since we may not consider ourselves to be clumsy, or we may guard against clumsy behavior ourselves if we ever go near a boat.

Therefore, social cognition may occur primarily to explain other people's behavior, satisfying a motive to make sense of that behavior. As a consequence, social cognitions are responsible for constructing our perceived social world. This world is constructed logically, using deductive and inductive reasoning to find solutions to *a* and *s* in an attempt to solve Why_t.

Magical Thinking and Anthropomorphism

Social cognition satisfies Why_t; it provides an answer to a why question regarding human behavior. Some philosophers now question whether this is necessary in order to predict behavior: they argue for a view of explanation as epiphenomenal to prediction, not dependent (Andrews, 2012). The reasoning behind this suggests that complex concepts necessary to generate explanations are not necessary in order to predict behavior. In support of this argument, research in the next chapter demonstrates that children can predict behavior before they have fully developed language abilities or complex abstract concepts. This line of argument suggests that explanation and prediction may function via separate mechanisms.

A functionalist account of explanation (for instance Putnam, 1960, 1967) suggests it could trigger social cognition; agents with mental states function or behave in the right way—just as a creature with mental states would function or behave. These behaviors include originating action, decision making, and so forth. This functionalist account can lead to errors because things that function in a particular way that do not have mental states can have mental states attributed to them.

People bring objects to life. Cars, computers, even brooms, all tend to elicit social cognition. This is actually quite remarkable since people preserve the notion that the object is indeed not a human being. Why is preservation of this notion necessary? The answer appears to be because people are governed by moral rules that prohibit harm. As a result, it is acceptable to shut down (or even destroy) a computer program that is beating you at chess, but it violates social convention to upset the chessboard if you are losing to a colleague, and it violates moral rules to attack said human opponent as she celebrated impeding victory. This singular difference suggests a dissociation between social cognition directed toward people and objects, and the literature supports such a distinction.

Social cognition recruits a brain network, including MPFC, temporal regions extending from the temporoparietal junction (TPJ) along the superior temporal sulcus (STS) to the anterior temporal pole, parietal regions including the precuneus and posterior cingulate, and subcortical regions such as the amygdala (Harris & Fiske, 2009). This reliable brain system engages for a variety of social cognitive tasks, including mental state inferences. For instance, dispositional attributions—a type of social cognition—engage parts of this brain network. If participants read that J.T. sings to a rain god, then read low consensus information about other people (hardly anyone else sings to a rain god), low distinctiveness information from other entities (J.T. sings to the sun god, the moon god, and the wind god as well), and high behavioral consistency information about another person (in the past, J.T. always sang to the rain god), they then make a dispositional attribution for the behavior (she must be responsible for her singing, perhaps she believes that deities exist) (Kelley, 1972; McArthur, 1972). Participants recruited their MPFC and STS to help them distinguish between the different base-rate information present in the sentences about J.T.'s behavior, with these regions becoming active when presented with the correct combination of consensus, distinctiveness, and consistency information (Harris, Todorov, & Fiske, 2005).

A replication of the above paradigm using objects engaged in action instead of people resulted in an overlapping yet different brain network (Harris & Fiske, 2008). When participants read that the pen fell off the table, hardly any other objects fell off the table, the pen falls off almost every other surface, and in the past, the pen would always fell off the table, they made dispositional attributions to the pen (perhaps the pen is responsible for the behavior; it must be alive or possessed!). However, these participants relied on their STS and amygdala, swapping the latter for the MPFC. This difference in activation may be sufficient to preserve the notion that the pen is not a human being and not deserving of protection. How might this thought process occur? The amygdala is involved in vigilance and detects emotionally salient occurrences and stimuli in the environment. Given that most pens do not fall off tables on their own, low consensus information about the behavior of the object may have engaged the amygdala as a threat detection or salience mechanism. Preserving engagement of the STS may ensure that the dispositional attribution is made, but failure to engage in the MPFC may restrict positive affect and application of moral rules and social conventions to the pen. The MPFC integrates information, performing relevant computations governing executive functions. Yet it is biased to respond to positive affect and reward, preferring social rewards in particular. Therefore, this crucial distinction in brain activation may result in a mental state inference bereft of the necessary affective components that mark an agent as human, thus deserving of protection.

Patient research also supports the role of the amygdala in anthropomorphism. The classic object anthropomorphism paradigm developed in the 1940s had participants view geometric shapes in nonrandom, agentic motion while they described what they saw, just as I described at the beginning of this chapter (Heider & Simmel, 1944). Most people described the behavior of the shapes as involving human or animal characteristics, including hunting, chasing, and fighting. People engage social cognition to the shapes in nonrandom motion. In comparison, a control group of subjects saw the same stimuli in random motion and failed to personify, anthropomorphize, or engage social cognition to the shapes to explain their apparent behavior. More recently, when a patient with bilateral amygdala damage viewed the same stimuli, she also failed to engage social cognition or to anthropomorphize, describing the apparent behavior in strictly geometric terms as illustrated in the excerpt at the beginning of the chapter (Heberlein & Adolphs, 2004). This suggests that the amygdala is necessary to perceive biological motion in objects and corroborates the brain-imaging evidence.

Dehumanizing People

People can view other people as less than human or engage in a dehumanized perception. This is not the other side of the anthropomorphism coin since dehumanized perception involves reduced activity in the social cognition brain network, particularly the MPFC (Harris & Fiske, 2006, 2007, 2011). In the initial experiments participants viewed pictures of traditionally dehumanized groups: extreme social out-groups, specifically stereotypic representations of homeless people and drug addicts, along with pictures of other people from a variety of positive and ambivalent social groups (such as business people, college students, disabled people). Traditionally dehumanized groups are viewed as low in trait warmth and competence and elicit disgust instead of another social emotion that requires social cognition. Pictures of such groups elicited significantly less social cognition brain network activity than pictures of other social groups (Harris & Fiske, 2006). However, having participants infer the food preference of the dehumanized people reengaged social cognition brain network activity, suggesting that participants could engage social cognition to these dehumanized groups (Harris & Fiske, 2007). Evidence that the dehumanized perception occurs spontaneously comes from an EEG study that demonstrates that people's brain waves over medial frontal cortex differentiate dehumanized and humanized groups after about 140 milliseconds (Harris & Fiske, 2009). Subsequent studies verified that this reduced brain network activity indexed less social cognition: participants also use fewer mental state verbs when asked to describe a day in the life of these people and rate them lower on human-related traits such as intelligence and articulateness (Harris & Fiske, 2011). Importantly, dehumanized perception as defined here results from social comparison: the extreme out-groups are dehumanized relative to other social groups.

Published the same year as the initial dehumanized perception brain-imaging paper, another brain-imaging paper showed a similar pattern of activity when experienced video-game players killed avatars of their friends in the context of a violent first-person shooter game (Mathiak & Weber, 2006). In this study participants assumed the role of a gunman, and the objective in this virtual reality was "kill or be killed." Reduced social cognition brain network activity occurred during violent exchanges compared to nonviolent virtual exchanges. In this study reduced social cognition brain network engagement even stretched to subcortical structures such as the amygdala. This study not only confirms a role for reduced social cognition in dehumanizing behavior but also suggests that a dehumanized perception actually may drive behavior.

Interestingly, dehumanization as a topic of theoretical inquiry has been associated with explanation. The theoretical arguments posit that perceiving victims of previous wrongdoings by the self or the in-group in a dehumanizing manner legitimizes those behaviors (Bar-Tal, 1990; Castano & Giner-Sorolla, 2006). Therefore, one function of dehumanized perception is explaining why bad things may have occurred, a post hoc justification: the victims were not human, thus suggesting a bad thing did not really happen, or the bad outcome was somehow deserved.

A second function of dehumanized perception may be as a proactive *emotion regulation strategy* (Cameron, Harris, & Payne, 2016; Harris & Fiske, 2009; Shaw, Batson, & Todd, 1994). People may predict that an impending social interaction may be unpleasant and resort to dehumanization as a way of regulating the emotions concurrent with such a prediction. In this way dehumanization may function in a similar manner to distraction or cognitive disregard (Rodin, 1987), allowing the observer to not attend to the presumed unpleasant mental states of the dehumanized other. This function of dehumanized perception achieves the same goals as reappraisal or suppression, other emotion regulation strategies. Dehumanized perception as emotion regulation may actually be necessary in some social contexts, suggesting a possible reason why it provided an evolutionary advantage and remains a feature of human cognition. I discuss specific social contexts that encourage dehumanized perception in chapters 8 through 10, but for the purposes of the current discussion, suffice it to say that a common feature of these social contexts is that dehumanized perception moderates emotional responses.

A third function of dehumanized perception related to its role as an emotion regulation strategy surrounds its usefulness in converting people to means to an end. Stated differently, dehumanized perception allows people to become instrumental (see Cikara, Farnsworth, Harris, & Fiske, 2010, for brain-imaging evidence). Imagine a situation in which a person is fleeing through a jungle from some assailant a few miles behind. Such a threat to survival suggests the person may resort to any opportunity to enhance her chances of survival, including using the physical environment to an advantage. In order to do so, the environment may require manipulation (the tearing of a branch from a tree to use as a weapon while lying in wait for the assailant). Similarly, the person may use the social environment to her advantage, manipulating people to ensure greater chances of survival (convincing nearby local village fishermen to join her as she lies in wait for the assailants). The behavior in the example could be interpreted as cooperative behavior, suggesting that if the person required a dehumanized perception to motivate the fishermen's helping behavior, then dehumanized perception might underlie more than just negative behavior.

The positive effects of dehumanized perception can be made clearer by another, real-world example. In 2007 a construction worker, while waiting on the busy New York City subway platform with his two young daughters, noticed a young man having a seizure. He and another passerby quickly attended to the young man; however, the violent convulsions caused the young man to fall onto the tracks in front of an oncoming train. In an instant the construction worker dove onto the tracks, dragged the young man into a nearby trench, and lay on top of him as the train passed overhead. Why would a person commit such a heroic act? The person being saved suffered from epilepsy, a mental illness. Therefore, it is possible that the hero did not infer the person's mental state because it was diminished, and this motivated the hero to react and save the person. At the moment of the seizure, mental state inferences may not have occurred, causing the default response to the social context that someone was in need of help, thereby motivating the behavior. Indeed, there is someone present in the description of the social context, and of course people are saved in part because they have mental states. But the thought in the moment preceding the helping behavior may not have been a mental state inference but, rather, a thought about the social context. Therefore, it is important to remember that the description of mental state inference is a flexible process, and it is possible to both infer and not infer the mental states of the same person in different social contexts. So diminished mental state inferences can lead to positive behavior: they are not always the causes of bad things. A mental state inference in our example context might have led to empathy, the experiencing of unpleasant feelings, and to diffusion of responsibility (see Darley & Latane, 1968; Latane & Darley, 1968). But perhaps the normative response bereft of a mental state inference led to the heroism. Similar effects are found in experiments in which I give participants descriptions of crimes, and I make the person's mental state less salient. In these studies people reduce responsibility attributions for the crime to the perpetrator and recommend less harsh punishment (Capestany & Harris, 2014).

But is a dehumanized perception really necessary to motivate someone to help? Does helping not result from compassion, empathy, and other more prosocial motives? Perhaps, but those motives result from explanation. For instance, if someone feels compassion toward a homeless person and is motivated to help, explanation presumably plays a role; the individual would have to justify that helping the person is consistent with being compassionate and that the homeless person is deserving of compassion, regardless of whether the homeless person is genuinely in need or a charlatan. This suggests that those motives need not even be accurate but only need to be believed by the person motivated to help. Therefore, purely positive motives can be manifested if someone is convinced that he or she ought to have those motives, even if those motives are not accurate responses to the social context (for instance, if the person fleeing to the village fishermen for help in the example above is an escaped criminal). In order to convince someone that she should adopt inaccurate motives, a dehumanized perception may first be necessary in order to manipulate her perception of the appropriate response to the social context. Stated differently, dehumanized perception may lie behind deception itself (I discuss this concept further in chapter 6).

This idea of the instrumental function of dehumanized perception is hinted at in research demonstrating that ambivalently sexist men engage the dehumanized perception brain activity pattern when looking at pictures of scantily clad women (Cikara, Eberhardt, & Fiske, 2011). Here, the relevant social comparison is pictures of non-scantily clad women. This study suggests that men who view women as people to be nurtured, cared for, put on a pedestal, but not as equals deserving of employment beyond the family and home, view pictures of women's bodies as objects or instruments. Other research using eye tracking has confirmed that such men are looking at the women's bodies in these pictures, not at their faces, which typically capture attention (Bernard, Gervais, Allen, Campomizzi, & Klein, 2012). Together, this research provides evidence of the objectification of women[1] and underlines the function of dehumanized perception in the instrumental use of people. I describe similar effects of dehumanized perception in later chapters.

The Curious Case of Animals

Finally, a brief word about animals. Not all animals are perceived equally. In fact, people reserve the social cognition brain network for interactions with other people, themselves, and perhaps pets. For all other animals, people tend not to engage all of their social cognitive processing. Certainly STS and temporal regions become engaged by animal parts such as faces (Chao, Martin, & Haxby 1999) as well as biological movement (for discussion, see Puce & Perrett, 2003). However, full engagement of the network tends not to be reserved for all animals. The curious case of pets, animals that share the physical and social environment of people, suggests that the brain's response to animals is malleable. The complaint with this explanation surrounds why the default remains processing with other brain systems. Animals are an important resource; although today many serve companionship and work roles, some animals have been used as food for humans throughout our evolutionary history. As a result, imbuing them with moral protection defeats this purpose. Instead, some animals have risen above this default, and some people have personally elevated all animals above this default; these cases perhaps result from engagement of the social cognition brain network. Certainly vegans and vegetarians such as Peter Singer (1980) argue that we violate moral rules by consuming animals, suggesting that people typically fail to engage the social cognition brain network for agents that they should. Similarly, most dog and cat lovers are convinced of their pet's personality or disposition, can communicate with their animals, and engage in social cognition (as described in chapter 2).

The central issue in this discussion is morality; attributing humanity provides moral protection. Researchers have discovered that people's propensity to eat meat is related to their inference of the mental state of animals; people who believe animals lack mental attributes and are highly dissimilar to humans are more likely to eat meat (Loughnan, Bastian, & Haslam, 2014). Therefore, people are able to solve the meat paradox—to simultaneously hold the belief that animals should be cared for, yet eaten—by denying mental states to some animals and thinking of such animals as dissimilar from humans. This illustrates wonderfully how explanation shapes behavior.

This attribution of social cognition to some animals, who are not more related genetically to humans than other animals such as other primates, has elevated these animals to a preserved status in human perception. The irony should not be lost that members of another species receive moral protection, while so many human beings are robbed of it as happens in cases of genocide and slavery, where dehumanization is hypothesized to be the active psychological mechanism. Perhaps a phenomenon like explanation is a fitting reason (although the survival benefits provided by dogs, for instance, should not be ignored).

Final Thoughts on Explanation

I have argued that social cognition solves a Why_t, satisfying the need for explanation. This can be clearly observed in the case of anthropomorphized agents and even when social cognition is withheld from people and they are dehumanized. Social cognition (or lack of spontaneous social cognition) provides information that can be used as a reason to explain the agent's behavior or behavior toward the agent. This is a powerful feature of social cognition because it implicates social cognition in the active construction of reality. Interestingly, social cognition occurs without regard to accuracy of cognitions but does occur with regard to consistency of cognitions. Switching accuracy for consistency creates different standards for Truth, suggesting that people's opinions of others can be firmly maintained even in the existence of refuting evidence. This feature of humanity may underlie intractable group and interpersonal conflict, suggesting that such misunderstandings can be solved by moving the focus from consistency to accuracy or at least by recognizing where the inconsistencies lie.

For instance, people involved in intractable conflicts tell a narrative that is consistent with the conflict. The narrative may often involve historical wrongs committed by the out-group, instances of their inhumane nature, and justification for the ongoing conflict. Such narratives require interpreting new actions by the out-group as being consistent with the narrative. Therefore, even reconciliatory behavior may not be interpreted as such because it is inconsistent with the narrative. Herein lies the problem with consistency. Instead, if the focus were on accuracy (perhaps mediated by a third party perceived as unbiased), then reconciliatory attempts would stand a better chance of being effective, and the narrative might slowly change. This is the hope in any case. Now on to naive scientists to examine whether social cognition abilities are learned or innate.

Note

1. Despite independent research histories, objectification and dehumanization theories today are united. Early research on objectification grew out of feminist psychology and focused on self-objectification of women. Early theories argued that an objectified gaze toward women could lead to women internalizing these perceptions, thus viewing themselves more as physical objects than as people (Fredrickson & Roberts, 1997). Subsequent research demonstrated many detrimental cognitive effects of self-objectification. A few years later, the first experimental studies of dehumanization focused on ascriptions of secondary (social) emotions to out-groups (Leyens, Rodriguez-Perez, Rodriguez-Torrez, Gaunt, Paladino, Vaes, et al., 2001); secondary emotions require mental state inferences. After about a decade of independence, the two literatures began to overlap, and today most researchers view objectification as a form of dehumanization, consistent with subsequent theory that describes dehumanization to animals and to objects (Haslam, 2006).

References

Andrews, K. (2012). *Do apes read minds? Toward a new folk psychology*. Cambridge, MA: MIT Press.

Bar-Tal, D. (1990). Causes and consequences of delegitimization: Models of conflict and ethnocentrism. *Journal of Social Issues*, 46(1), 65–81.

Bernard, P., Gervais, S. J., Allen, J., Campomizzi, S., & Klein, O. (2012). Integrating sexual objectification with object versus person recognition: The sexualized-body-inversion hypothesis. *Psychological Science*, 23(5), 469–471.

Cameron, C. D., Harris, L. T., & Payne, B. K. (2016). The emotional cost of humanity: Anticipated exhaustion motivates dehumanization. *Social Psychological & Personality Science*, 7, 105–112. doi:10.1177/1948550615604453.

Capestany, B. H., & Harris, L. T. (2014). Disgust and biological descriptions bias logical reasoning during legal decision-making. *Social Neuroscience*, 9, 265–277. doi:10.1080/17470919.2014.892531.

Castano, E., & Giner-Sorolla, R. (2006). Not quite human: Infrahumanization in response to collective responsibility for intergroup killing. *Journal of Personality and Social Psychology*, 90(5), 804.

Chao, L. L., Martin, A., & Haxby, J. V. (1999). Are face-responsive regions selective only for faces? *Neuroreport*, 10(14), 2945–2950.

Cikara, M., Eberhardt, J. L., & Fiske, S. T. (2011). From agents to objects: Sexist attitudes and neural responses to sexualized targets. *Journal of Cognitive Neuroscience*, 23(3), 540–551.

Cikara, M., Farnsworth, R. A., Harris, L. T., & Fiske, S. T. (2010). On the wrong side of the trolley track: Neural correlates of relative social valuation. *Social Cognitive and Affective Neuroscience*, 5, 404–413.

Darley, J. M., & Latane, B. (1968). Bystander intervention in emergencies: Diffusion of responsibility. *Journal of Personality and Social Psychology*, 8(4), 377.

Fredrickson, B. L., & Roberts, T. A. (1997). Objectification theory. *Psychology of Women Quarterly*, 21(2), 173–206.

Harris, L. T., & Fiske, S. T. (2006). Dehumanizing the lowest of the low: Neuro-imaging responses to extreme outgroups. *Psychological Science*, 17, 847–853.

Harris, L. T., & Fiske, S. T. (2007). Social groups that elicit disgust are differentially processed in the mPFC. *Social Cognitive and Affective Neuroscience*, 2, 45–51.

Harris, L. T., & Fiske, S. T. (2008). Brooms in *Fantasia*: Neural correlates of anthropomorphizing objects. *Social Cognition*, 26, 209–222.

Harris, L. T., & Fiske, S. T. (2009). Social neuroscience evidence for dehumanised perception. *European Review of Social Psychology*, 20, 192–231.

Harris, L. T., & Fiske, S. T. (2011). Dehumanised perception: A psychological means to facilitate atrocities, torture, and genocide? *Zeitschrift fur Psychologie. Journal of Psychology*, 219, 175–181.

Harris, L. T., Todorov, A., & Fiske, S. T. (2005). Attributions on the brain: Neuro-imaging dispositional inferences, beyond Theory of Mind. *NeuroImage*, 28(4), 763–769.

Haslam, N. (2006). Dehumanization: An integrative review. *Personality and Social Psychology Review*, 10(3), 252–264.

Heberlein, A. S., & Adolphs, R. (2004). Impaired spontaneous anthropomorphizing despite intact perception and social knowledge. *Proceedings of the National Academy of Sciences of the United States of America*, 101(19), 7487–7491.

Heider, F., & Simmel, M. (1944). An experimental study of apparent behavior. *American Journal of Psychology*, 57, 243–259.

Kelley, H. H. (1972). Attribution in social interaction. In E. E. Jones, D. E. Kanouse, H. H. Kelley, R. E. Nisbett, S. Valins, & B. Weiner (Eds.), *Attribution: Perceiving the cause of behavior* (pp. 1–26). Hillsdale, NJ: Lawrence Erlbaum Associates.

Latane, B., & Darley, J. M. (1968). Group inhibition of bystander intervention in emergencies. *Journal of Personality and Social Psychology*, 10(3), 215.

Leary, M. R. (2004). *The curse of the self: Self-awareness, egotism, and the quality of human life*. New York: Oxford University Press.

Leyens, J. P., Rodriguez-Perez, A., Rodriguez-Torres, R., Gaunt, R., Paladino, M. P., Vaes, J., et al. (2001). Psychological essentialism and the differential attribution of uniquely human emotions to ingroups and outgroups. *European Journal of Social Psychology*, 31(4), 395–411.

Loughnan, S., Bastian, B., & Haslam, N. (2014). The psychology of eating animals. *Current Directions in Psychological Science*, 23(2), 104–108.

Mathiak, K., & Weber, R. (2006). Toward brain correlates of natural behavior: fMRI during violent video games. *Human Brain Mapping*, 27(12), 948–956.

McArthur, L. Z. (1972). The how and what of why: Some determinants and consequences of causal attribution. *Journal of Personality and Social Psychology*, 22, 171–193.

North, M. S., & Fiske, S. T. (2012). A history of social cognition. In A. W. Kruglanski, & W. Stroebe (Eds.), *Handbook of the history of social psychology* (pp. 81–99). New York: Psychology Press.

Puce, A., & Perrett, D. (2003). Electrophysiology and brain imaging of biological motion. *Philosophical Transactions of the Royal Society of London. Series B, Biological Sciences*, 358(1431), 435–445.

Putnam, H. (1960/1975). *Minds and machines, reprinted as Mind, language, and reality* (pp. 362–385). Cambridge, UK: Cambridge University Press.

Putnam, H. (1967/1975). *The nature of mental states, reprinted as Mind, language, and reality* (pp. 429–440). Cambridge, UK: Cambridge University Press.

Rodin, M. J. (1987). Who is memorable to whom: A study of cognitive disregard. *Social Cognition*, 5(2), 144.

Shaw, L. L., Batson, C. D., & Todd, R. M. (1994). Empathy avoidance: Forestalling feeling for another in order to escape the motivational consequences. *Journal of Personality and Social Psychology*, 67, 879–887. doi:10.1037/0022-3514.67.5.879.

Singer, P. (1980). Utilitarianism and vegetarianism. *Philosophy & Public Affairs*, 9(4), 325–337.

Waytz, A., Morewedge, C. K., Epley, N., Monteleone, G., Gao, J., & Cacioppo, J. T. (2010). Making sense by making sentient: Effectance motivation increases anthropomorphism. *Journal of Personality and Social Psychology*, 99(3), 410–435. doi:10.1037/a0020240.

5 Naive Scientists

Two balls are rolling. The first ball suddenly becomes stationary. The second ball changes its direction of movement and now moves toward the first ball and hits it. The first ball immediately starts to roll upon contact with the second ball.

The first ball sits stationary. A second ball rolls along and hits the first ball. The first ball immediately starts to roll upon contact with the second ball.

The first ball sits stationary. A second ball rolls along and hits the first ball. After a delay of a few seconds, the first ball starts to roll upon contact with the second ball.

The scenarios described above are inspired by paradigms in developmental psychology that attempt to determine whether nonverbal infants attribute intentionality and causality. Did the second ball intend for the first ball to roll, did it cause the action, or was it coincidence? Infants have been solving puzzles like these for well over fifty years in developmental psychology labs. The results suggest something quite remarkable about human cognition: it has the ability to determine agency. Agency is an important concept because it identifies objects in the environment that initiate their own behavior (Fiske & Taylor, 2013). Many objects in the physical environment can be manipulated or acted upon; however, some objects also initiate behavior without any external force acting upon them. Differentiating these two types of objects is crucial for survival, and it suggests a basic platform on which social cognition might be built (see figure 5.1). Certainly, the developmental research seems to suggest that the ability to differentiate agents[1] from nonagents arises early in life. Moreover, this infant research suggests that the mechanism utilized for such a task is based on logic.

Despite popular accounts to the contrary, human beings employ widespread use of logic and reason. From the ancient Egyptians and Greeks through the Renaissance into modern capitalist societies, human beings have prized logic and reason as both desirable abilities and a combined approach to solving conflicts, disputes, and problems; logic and reason, it has been argued, should be used to govern society (Pinker, 2011). Interestingly, human beings use logic and reason not only to solve

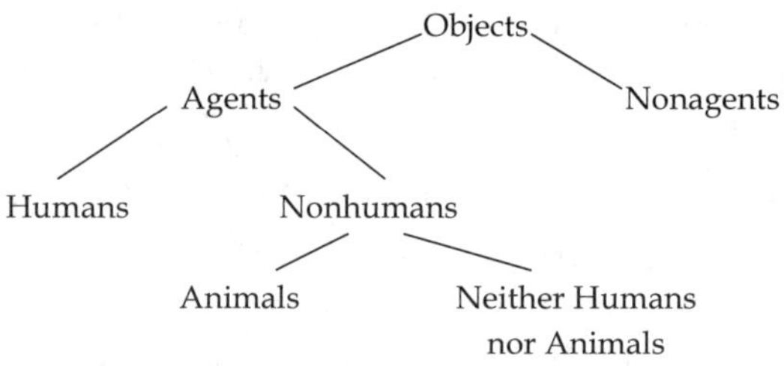

Figure 5.1
A diagram of possible perceptions of different types of objects, meant to stir discussion and scientific debate. Each branch of the tree represents a dissociation. This toy model is one hypothesis for how infants may perceive the world, and in their perception, objects are initially all things in physical environment, including humans. It does not mean humans are really a type of object. Children first dissociate agents from nonagents before dissociating humans from nonhumans. Presumably, they then dissociate animals from both humans and nonanimals.

societal problems but also to learn about other people and the social world. Each human being can be thought of as a *naive scientist* (Heider, 1958) since she uses scientific, statistical principles to understand other people and to infer the contents of their minds. Following scientific principles, an inference—causal or otherwise—is made with a certain degree of probability given base rates of occurrences and their variation. Similarly, people make mental state inferences and formulate stereotypes about groups based on statistical occurrences within the variability of behavior.

The naive science term applies well to infants who are constantly hypothesis testing. Like any good scientist, the infant requires some theoretical background to generate hypotheses and interpret statistical evidence. Interestingly, in the domain of social cognition, developmental psychologists have suggested that change in movement is sufficient to dissociate two types of objects: those that are self-propelled and those that are not (Premack, 1990). This distinction maps on nicely to the human–nonhuman appraisal that triggers social cognition and suggests an innate foundation on which social cognition may be built. The developmental theorists argue that carving the world up into two types of objects allows the infant to apply distinct concepts to agents. If a first non-self-propelled object's motion is changed by a second object, the infant attributes causality for the change in motion to the second object. If a self-propelled object exhibits change in motion, then the infant attributes intentionality to the self-propelled object. Finally, if two self-propelled objects are related in a special way, then goals can be attributed to one of the objects. Researchers argue that causality, intentionality, and goal attribution then are hard-wired concepts innate to human beings. If this theory is true, it suggests that intentionality and goal attribution are basic building blocks for social cognition.

This idea of innate principles that guide intentionality and causal attribution is similar to a naive theory of physics (Spelke, 1990) that infants also possess about the physical world (knowing not to crawl off a precipice). These rules are not taught, but they are early understandings (without experience or teaching) that promote survival. Similarly, in the social domain, dissociating agents from nonagents may reflect a survival advantage. Interestingly, this social distinction also depends on changes in the perceptual field regarding movement.

Neurons in the temporal lobe, specifically the superior temporal sulcus (STS), track biological movement and dissociate it from nonbiological movement. Together, these findings suggest that movement is the key feature on which social cognition is scaffolded. Detecting changes in movement is a relevant survival skill since predators and other agents that initiate their own action may decide to initiate harmful behavior. Therefore, it is in the organism's best interest to dissociate agents from nonagents and biological motion from nonbiological motion. In support of this, evidence suggests that infants as young as six months have a different expectation for the interaction of objects with human hands versus an inanimate object (Leslie, 1982, 1988). Ten-month-olds expect to influence animate objects, but not inanimate objects, from a distance (Golinkoff, Harding, Carlson, & Sexton, 1984).

Infants Dissociate People from Objects

Perceptual information, however, is not sufficient to dissociate intentionality. Instead, infants rely on the equifinal structure of actions—in different social contexts, the agent's actions result in the same consequence (Heider, 1958). Therefore, tagging an agent may result in a dispositional inference to the agent over time. This then results in a prediction (often incorrect) of the agent's goals based on rational, logical reasoning (Dennett, 1987). Only then, it is argued, can intentionality be inferred. This intentional stance situates intentionality as complex social cognition akin to dispositional inferences. A simpler definition of intentionality stems from considering the inference of agency as an inference of rational agency—the default may be to view all behavior as purposeful, intentional, and rational given the social context and the agent's beliefs about the social context (Gergely, Nádasdy, Csibra, & Biro, 1995).

A test of this inherent rational agent concept (referred to as the teleological stance) showed infants an animation with two circles and a rectangle, the latter positioned between the former (Gergely et al., 1995). Agency was attributed to the circles because they both expanded and contracted in size in sequence. After this exchange, one circle approached the other, stopping at the rectangle. The circle then rolled away from the obstacle presented by the stationary rectangle before rolling toward the rectangle once again and jumping over the rectangle. On reaching the second circle, the animated circles resumed their expansion–contraction exchange. This video was compared to an identical video in which the rectangle was now positioned behind the first circle instead of between the two circles. Now the jumping behavior of the first circle seemed irrational since it is not the shortest way to get to the second circle. To test whether infants attributed rational action to the circle, they were given two test events to witness in which, without the rectangle, the first circle either went in a straight line to the second circle or performed the jumping motion on the way to the second circle. In the first condition where the infants observed the rational obstacle jumping, they looked longer at (dishabituated to) the straight line test event more than at the jumping test event. However, in the second condition, where infants saw the nonrational jumping, they did not differentially dishabituate to the test events. This study suggested that rationality is a fundamental part of agency attribution, and it has led researchers to argue that social cognition can result from teleological reasoning (Csibra & Gergely, 1998).

Other relevant work comes from studies that describe moral judgments and actions in preverbal infants and toddlers (see Hamlin, 2013, for a review). Across a number of studies, researchers describe moral evaluations from children with limited moral learning, suggesting that the ability to have sophisticated moral evaluations may be innate. In one such study a puppet displays a preference for one toy over another (Hamlin, Ullman, Tenenbaum, Goodman, & Baker, 2013). Two other puppets witness the first puppet's behavior, before both toys are removed and placed behind doors. One of the observer puppets then lifts the door concealing the preferred toy, while the other lifts the door concealing the nonpreferred toy. Ten-month-old infants prefer the puppet that revealed the preferred concealed toy to the puppet that revealed the nonpreferred concealed toy. This very sophisticated moral evaluation of the puppets suggests that these infants were able to infer that one puppet was helping, whereas the other was not. However, such results (see also Hamlin, Wynn, & Bloom, 2007) are still hotly debated in the field.

But none of the research discussed thus far has demonstrated that infants can actually distinguish between humans and nonhumans. Certainly, an ability to distinguish agency and moral from immoral agents is useful, but humans do not exclusively belong to the category of agents. Certainly Disney has emphatically ensured as much. However, evidence does exist that infants respond differently to humans compared to nonhumans. In an illustrative study researchers habituated five- to nine-month-old infants to a human hand reaching for and grasping either a near or far object (Woodward, 1998). After habituation, a new video showed the same hand reaching for either the same object or the object in the first object's position (near or far). In another condition researchers repeated this sequence, replacing only the human hand in the video with a rod with bits of paper on the end to imitate an arm and fingers. Infants attended to the new object being grasped (in the old position) rather than the hand path-of-motion (grasping the old object in the new position), but only in the human hand, not the rod and paper condition. This study suggests that nine-month-old infants are sensitive to human goals to some degree and dissociate human from nonhuman action, even apparently agentic action. Subsequent studies replicated these effects with five- and six-month-olds and ruled out occlusion by the hand or movement of the fingers as alternate hypotheses (Woodward, 1998).

Fascinatingly, this human–nonhuman distinction holds even with agentic nonhumans. These longitudinal studies track infants looking, smiling, vocalizing, and moving their arms toward their mothers, a female stranger, and a doll, each active and animate or not (Legerstee, Pomerleau, Malcuit, & Feider, 1987). Although tracking began at three weeks, at five weeks infants began showing differential behavior toward the human and nonhuman agents. Infants look more, show more positive vocalizations and fewer negative vocalizations, reach toward, and smile toward humans versus the doll despite maintaining a distinction between animate and inanimate agents throughout the course of the age span. Furthermore, the infant builds on the initial human–doll distinction to further dissociate the mother from the female stranger. There was also an interaction, such that agentic humans always produced more smiling than nonagentic humans and dolls until fifteen weeks, and they started smiling more at their agentic mother exclusively at forty weeks (Legerstee, et al., 1987). Similar results were obtained when a doll was replaced by an object meant to approximate a face (eyes, nose, mouth, hairline) (Ellsworth, Muir, & Hains, 1993). These data suggest that although a strict human–nonhuman distinction does not emerge in the first few weeks of life, it may be present when the infant is considering agents from birth. This provides further evidence that this distinction may be innate (or occur very early in life). Again, such results are hotly debated and the jury is still out.

Interestingly, the interaction results described above seem similar to interaction patterns observed in the medial prefrontal cortex (MPFC) when considering human–nonhuman distinctions. For instance, in the studies on social reward described in chapter 4, the MPFC showed interaction effects between positive reward and human–nonhuman stimuli. Positive reward preferences, like agency, seem to be the default. Regions in MPFC reflected positive–negative distinctions as well as human–nonhuman distinctions, whereas other regions reflected the interaction with maximal activity for positive social stimuli (Harris, McClure, Van den Bos, Cohen, & Fiske, 2007; Van den Bos, McClure, Harris, Fiske, & Cohen, 2007). Perhaps these brain activity patterns reflect a basic principle within the governance of humans as stimuli: responses to human stimuli build on innate biases or preferences for movement and positive affect, exacerbating the response to these and other variables that promote survival and reproduction. Further developmental theory supports this idea, demonstrating that the ability to discern face-like features from objects may be innate, but the ability to discern conspecifics does not emerge in humans until age two months (Morton & Johnson, 1991). This suggests that the ability to discern face-like structure becomes mapped onto conspecifics, and it provides further evidence in support of the additive scaffolding of social cognition.

Language, Intention, and Emotion

I have talked before (see chapters 2 and 4) about the importance of language in social cognition, suggesting that the explanatory function of this ability resolves the tension to understand behavior. Beyond simply attentional and other movement-related differences (e.g., grasping, smiling) more associated with behavioral output, do infants also dissociate human–nonhuman stimuli with language? Such experiments have not yet been completed, but researchers have discovered that infants depend on the human–nonhuman distinction to learn language. The burgeoning language abilities of sixteen- to nineteen-month-olds is such that their already associated labels with toys allow them to participate in the achievement of joint reference, identifying discrepancies between an experimenter's focus and labels (Baldwin, 1991). Other evidence demonstrates that children rely on social cues to learn language; two-year-olds can associate labels with actions. In these studies children were able to learn words for intentional, not accidental, actions and for desired, not rejected, objects (Tomasello & Barton, 1994). Together, these studies suggest that infants can detect intentionality sometime between sixteen and twenty-four months, and they use this ability to guide word learning.

Related evidence describes the ability of eighteen-month-olds to infer intentionality, even without observing the completion of a goal. In these studies infants witnessed an adult attempt but fail, or attempt and succeed, to accomplish a goal by manipulating an object. For instance, one group of infants would witness an adult attempt to poke a button to activate a buzzer in a box with a stick, but occasionally miss getting the stick through the hole in the box and, when successful in getting the stick through, miss the button, failing to activate the buzzer. Other groups of infants would witness the adult interacting with the button, stick, and box or were simply allowed to interact with the objects themselves. Although none of these groups viewed the actual goal-directed behavior, only the failed attempt group re-enacted behavior. Stated differently, infants were able to discern the goal of the adults, re-enacting the successful behavior of activating the buzzer by poking the button with a stick through the hole in the box, even if they had never witnessed the correct behavior before. However, this effect broke down when a mechanical device performed the same movements as the human hand (Meltzoff, 1995). This finding provides further evidence of the human–nonhuman distinction and extends this distinction beyond language and agency to goals and intentions.

Theoretical Debate

Until this point, I have implicitly sided with the naive scientist account of how infants infer mental states. The broader theoretical debate surrounds the substance of mental state inferences. Is there a theory of mental states applied by people as naive scientists (Gopnik & Wellman, 1992; Heider, 1958), or do people simulate their own mental states in order to imagine what others' might be? This debate between (respectively) theory-theory and simulation-theory is old and well worn, yet neither theory provides a satisfying account. Theory-theory suggests that infants have a theory of behavior and their relation to mental states that they apply to each individual. However, as discussed above, it is possible to behave like an agent with a complex mental representation (a theory) without actually having a complex mental representation, something infants do not have. And so, theory-theory is an unsatisfying account. Simulation-theory suggests that infants simulate the mental states of others by imagining themselves in particular situations or performing particular behaviors; they thereby infer and understand the mental states of others. This account is unsatisfying as well since it is not possible to simulate experiences without having the experience oneself, and experience is something infants have very little of.

More recent theorizing suggests an important caveat in both cases and proposes a novel third theory to explain the substance of mental state inferences. First, one must consider that it is not necessary to have a complex representation of a concept (such as a mental state) in order to perform behavior consistent with mental state explanations. This depends on a distinction between prediction and explanation, which I have already adopted and is (partially) verified by neuroscience evidence. Therefore, many behaviors that infants perform may be masquerading as evidence of mental state inferences without the infants actually ever making such an inference. The critique against simulation theory is simply that infants have not had a wide array of experiences from which to simulate. Both critiques suggest restricted social cognition abilities in infants, which run counter to evidence presented above. Instead, the novel theory argues for a more social psychological account that factors in the behavior of people in situations (see Andrews, 2012). As we shall see, this call to arms for social psychology is occurring in the theory of other disciplines as well (see details of a similar occurrence in economics in chapter 8). Instead of formulating theories that depend on abstract concepts or simulating experiences one might never have had, it is simply necessary to know what people do in particular situations.

In a sense this novel theory is a revised theory-theory, since the hypothesis-testing element of theory-theory is preserved. Specifically, theory-theory states that infants, like scientists, generate hypotheses about behavior and underlying mental states and test those assumptions. Similarly, the novel theory provides a testable assumption but an assumption about the social context or situation, not the mental state or behavior. In young infants, the dissociation between agentic and nonagentic actors is sufficient to provide a testable hypothesis because agents are expected to exhibit rational behavior. Therefore, this innate distinction may be scaffolded upon, as what is considered rational behavior is expanded beyond the agency distinction to the human–nonhuman distinction. Even the definition of rationality may be refined, including consensus behavior and eventually social and moral norms or role-specific behavior. Thus, it may now be possible to predict and explain people's behavior without relying on abstract concepts or experience. Therefore, the human–nonhuman distinction we have described may be the core, fundamental process on which social cognition is built.

A similar attempt at resolving the theory-theory simulation theory debate argues for a minimal theory of mind (Butterfill &

Apperley, 2013). This argument hinges on encountering and registering as proxies for perception and belief respectively. Stated differently, an agent has experience (an encounter) that it records in a particular location (registers). This minimal theory frees the agent from abstract psychological concepts and may be considered a teleological approach to social cognition. Nonetheless, it shares a feature of the social-psychological account described above: it removes a reliance on abstract psychological concepts and focuses on the behavior of the agent in a specific context. Next, I describe the impact of the social context in more detail.

Note

1. An agent is any entity that seems to originate its own behavior. For instance, people may perceive plants as nonagents because they do not seem to originate their own behavior, often growing in the direction of sunlight. Moreover, the plant-initiated movement—growth—is so slow that on the human perceptual time scale, such movement is not detected moment to moment; plant growth is only noticed on the order of days or weeks. Hence, most plants are not attributed agency despite the fact that they are living organisms, adapting to survive on the planet and reproducing. However, a rock may be perceived as agentic if its movement seems to be nonrandom. If a boulder were rolling down a hill and then suddenly changed its direction, turning toward the nearby village, most people might attribute agency to the boulder. Therefore, agency attribution is independent of the biological underpinnings of the entity; it is a perceptual phenomenon based almost exclusively on movement. Bear in mind that this discussion is from the perspective of the infant, consistent with the chart in figure 5.1. Indeed, some nonagents may cause death, but by definition, nonagents cannot motivate their movement themselves; therefore, it may be best to attend to things that can, agents, for survival.

References

Andrews, K. (2012). *Do apes read minds? Toward a new folk psychology*. Cambridge, MA: MIT Press.

Baldwin, D. A. (1991). Infants' contribution to the achievement of joint reference. *Child Development*, 62(5), 874–890.

Butterfill, S. A., & Apperly, I. A. (2013). How to construct a minimal theory of mind. *Mind & Language*, 28(5), 606–637.

Dennett, D. C. (1987). *The intentional stance*. Cambridge, MA: Bradford Books/The MIT Press.

Fiske, S. T., & Taylor, S. E. (2013). *Social cognition: From brains to culture. Atlanta, GA: Sage Publications.*

Casasola, M., & Wilbourn, M. P. (2004). Fourteen-month-old infants form novel word-spatial relation associations. *Infancy*, 6, 385–396.

Csibra, G., & Gergely, G. (1998). The teleological origins of mentalistic action explanations: A developmental hypothesis. *Developmental Science*, 1(2), 255–259.

Ellsworth, C. P., Muir, D. W., & Hains, S. M. (1993). Social competence and person-object differentiation: An analysis of the still-face effect. *Developmental Psychology*, 29(1), 63.

Gergely, G., Nádasdy, Z., Csibra, G., & Biro, S. (1995). Taking the intentional stance at 12 months of age. *Cognition*, 56(2), 165–193.

Golinkoff, R. M., Harding, C. G., Carlson, V., & Sexton, M. E. (1984). The infant's perception of causal events: The distinction between animate and inanimate objects. *Advances in Infancy Research*, 3, 145–151.

Gopnik, A., & Wellman, H. M. (1992). Why the child's theory of mind really is a theory. *Mind & Language*, 7(1-2), 145–171.

Hamlin, J. K. (2013). Failed attempts to help and harm: Intention versus outcome in preverbal infants' social evaluations. *Cognition*, 128(3), 451–474.

Hamlin, J. K., Ullman, T., Tenenbaum, J., Goodman, N., & Baker, C. (2013). The mentalistic basis of core social cognition: Experiments in preverbal infants and a computational model. *Developmental Science*, 16(2), 209–226.

Hamlin, J. K., Wynn, K., & Bloom, P. (2007). Social evaluation by preverbal infants. *Nature*, 450(7169), 557–559.

Harris, L. T., McClure, S. M., Van den Bos, W., Cohen, J. D., & Fiske, S. T. (2007). Regions of the MPFC differentially tuned to social and nonsocial affective evaluation. *Cognitive, Affective & Behavioral Neuroscience*, 7(4), 309–316.

Heider, F. (1958). *The psychology of interpersonal relations*. Hove, UK: Psychology Press.

Legerstee, M., Pomerleau, A., Malcuit, G., & Feider, H. (1987). The development of infants' responses to people and a doll: Implications for research in communication. *Infant Behavior and Development*, 10(1), 81–95.

Leslie, A. M. (1982). The perception of causality in infants. *Perception*, 11(2), 173–186. doi:10.1068/p110173.

Leslie, A. M. (1988). The necessity of illusion: Perception and thought in infancy. In L. Weiskrantz (Ed.), *Thought without language* (pp. 185–210). New York: Clarendon Press/Oxford University Press.

Meltzoff, A. N. (1995). Understanding the intentions of others: Re-enactment of intended acts by 18-month-old children. *Developmental Psychology*, 31(5), 838.

Morton, J., & Johnson, M. H. (1991). CONSPEC and CONLERN: A two-process theory of infant face recognition. *Psychological Review*, 98(2), 164.

Pinker, S. (2011). *The better angels of our nature: The decline of violence in history and its causes*. London: Penguin.

Premack, D. (1990). The infant's theory of self-propelled objects. *Cognition*, 36(1), 1–16.

Spelke, E. S. (1990). Principles of object perception. *Cognitive Science*, 14(1), 29–56. doi:10.1207/s15516709cog1401_3.

Tomasello, M., & Barton, M. E. (1994). Learning words in nonostensive contexts. *Developmental Psychology*, 30(5), 639.

Van den Bos, W., McClure, S. M., Harris, L. T., Fiske, S. T., & Cohen, J. D. (2007). Dissociating affective evaluation and social cognitive processes in the ventral medial prefrontal cortex. *Cognitive, Affective & Behavioral Neuroscience*, 7(4), 337–346.

Woodward, A. L. (1998). Infants selectively encode the goal object of an actor's reach. *Cognition*, 69(1), 1–34.

6 The Delayed Sudden Death Virus Outbreak

Imagine a delayed sudden death virus outbreak in which a large proportion of the people on the planet become infected, and the others are not. Infection results in death, but not instantly: sudden death comes approximately one year after contracting the virus. However, there are no visible symptoms when a person is infected; infected persons simply know when they are infected because they have the thought in their mind, and they expect to die. The infection is spread by sharing the thought with another person in conversation, causing that person to have the thought. This disease has caused the governments of the world to institute a common live policy; people live in small groups of ten to twenty to pool resources and increase cooperation. Given that the infected people are constantly dying, there is migration between groups as suddenly surviving group members whose numbers have dwindled must now find new groups or recruit new members.

In the thought experiment above, *how do you determine whether a person who wants to join your group is infected, which would risk spreading the virus to you and your social group, or not?* This question is difficult because of the humans' ability to create false realities: to deceive. Since mental state inferences are inferences, not accurate perceptions, people must make decisions about the validity of those inferences. Moreover, they must also keep in mind the fact that other people may intentionally lead them to make false inferences. This latter wrinkle complicates mental state inferences. As described in chapters 1 and 3, people may assume that others are trustworthy by default. However, when this default assumption is questioned, people must now somehow decide what information is reliable in order to make the inference. I argue that the flexible social cognition helps to solve this problem.

To illustrate, let us consider a scenario I described at the beginning of the prologue: three assailants kill a person sitting at a street corner. How does a person get himself in a position where three people descend on him delivering demise? Certainly such a crime is premeditated, which suggests that the victim had previous social interactions with his killers or their employers. At worst, three assailants suggest a social group, so the victim was either being expelled from the social group or was perhaps a member of a competitive outgroup. As described in chapter 3, death by social group has always been a significant risk factor throughout human evolution and history due to the highly social nature of primates, especially the human animal. This is not a trait unique to people, but it is more commonly found in highly competitive groups in a variety of species. Nonetheless, death by social group seems prevalent in human evolution and history.

Given that encounters with agents are capable of causing death, the ability to distinguish agents from nonagents is acquired early in life and may even be innate. However, given that death by social group is also a significant additional risk factor, and human beings belong to social groups capable of causing death, human beings needed to distinguish other human agents from nonhuman agents very quickly. The developmental psychology research discussed in chapter 5 suggests that, after a few months, infants dissociate people from other nonhuman agents, supporting the idea that evolution prioritized human–nonhuman detection.

But not all agents or humans harm; some help. What is remarkable about humans is the compassion they show to others, often a distant trait in other, highly competitive mammals (although as discussed in chapter 3, this ability to show compassion may yet be discovered in these species). Since agents originate their own behavior, they are capable of causing harm, or of helping, based on their own desires, wishes, and intentions—in short, their mental states. As a result, it is essential to dissociate the helpers from the harmers within the category human. But as I mentioned, unlike most other agents, human beings can deceive; that is, they can present a false impression of themselves and their social context, leading a perceiver to draw erroneous explanations for their behavior and to make poor predictions when interacting with such people.[1] Perhaps this quandary is solved if social cognition can determine what category a possible human being may belong to: helper or harmer, friend or foe. I argue that implicit in the detection of another agent as a human comes a categorization of helper (or harmer) that guides the interaction. This is similar to the rational agent perception I described in chapter 5: if what is perceived is an agent, it must behave rationally, and things that behave rationally are agents. Similarly, if it is human, it must be helpful (or harmful), and if it is perceived as helpful (or harmful), it must be perceived as human.

For instance, consider the case of the tourist who visits a country for the first time on vacation. Such a person may assume that every person she meets, from checking in at the airport, checking in at the hotel, visiting the local restaurants and sites, is there to help her. This would not be a faulty assumption if indeed all of the people in the various service roles adhered to their roles and helped the tourist. Now consider the case of the terrorist who visits a country for the first time with the malignant intent to blow something up. Such a person may assume that every person she meets, from people at the airport, to people on the street, is there to thwart the plot. This would not be a faulty assumption if each of these people knew the true intention of the terrorist and did not mistake her for an eager tourist. These examples not only illustrate the default assumptions but highlight how the assumption can either be helper or harmer based on the goals, intentions, and other features of the agent (tourist or terrorist, *a* in the mental tension equation) as well as the social context (vacation or terrorist plot, *s* in the mental tension equation) that frames every interaction with another person.

Furthermore, this implicit categorization process may be guided by perceived social group membership. The tourist on vacation may categorize the person who picks her up from the airport as a thief, leading to a perception of that person as harmer, not helper. Similarly, the terrorist may categorize the person who picks her up at the airport as a comrade, leading to a perception of that person as a helper, not harmer. Such categorizations provide additional information beyond *a* and *s* that changes the outcome of the equation, resulting in different behaviors. Therefore, social groups are key to further understanding who is considered a helper and who a harmer.

The Self and Social Groups

There are huge advantages to social group formation. Cooperative, reciprocal behavior certainly allows the individual to gain immense benefits that could not otherwise be accrued (Fiske, 1992). However, difficult environmental contexts encourage competitive behavior over limited resources between and within groups. These conditions prioritize much baser survival and reproduction instincts. Instincts related to social life, such as cooperation, prosociality, empathy, and alliances become secondary. Therefore, the risk of death by social group is a simple condition of social life. Evolution perhaps preserved flexible social cognition as a trait to promote survival within social groups, allowing people to pursue secondary, social instincts while detecting those seeking to harm. But how does having an innate ability to distinguish between humans and nonhumans relate to dissociating harmful and helpful people (or deceitful from not deceitful people)? As just discussed, people may assume, by default, that other people will also be driven by these secondary, social instincts. Therefore, categorizing an agent as human automatically places the individual in the helper, friend, in-group, positive category (or harmer, enemy, out-group, negative category). Evidence for this default perception as positive comes from the developmental literature, where children show more interest, smiling, gestures, and other more positive behaviors to other people compared to nonhuman agents (see chapter 5). In this manner flexible social cognition aids in distinguishing humans from nonhumans, letting us know who is friend and foe, who is for us and who is against us.

Although I am agnostic as to whether helper or harmer is the default category when categorizing an agent as human, I think that helper may be the default for most people for a couple of additional reasons. First, humans evolved in small, relatively segregated social groups, suggesting that the majority of other humans that they encountered belonged to that in-group and thus were potential helpers. Second, people trust other people as a default response and respond to violations of trust harshly, displaying a negativity bias. However, it is possible that the default may change from person to person, whereby some people may perceive others as harmers by default. This may be due to psychological illness, past experiences, or a personality trait. Moreover, although most people would categorize Adolph Hitler as a human (even though some people may believe that Nazis were aliens who came to Earth to terrorize humanity and, as such, were not human), they may view him as a harmer, suggesting the two perceptions are orthogonal. Such instances aside, I still think that the default may be human equals helper.

Most people have described pro- and antisocial behavior in terms of basic survival, with a focus on the self and group behavior (for instance, see Dovidio, Gaertner, & Saguy, 2008). However, there is an inferential leap that occurs when people attempt to explain kin–in-group member based prosocial helping behavior and stranger–out-group member based on antisocial behavior as an extension of a self–other distinction. In this explanation proponents argue that the distinction between the self and others is blurred, allowing in-group members to be incorporated into the category of the self. This incorporation of the other with the self involves the perception of the other as similar to the self in some ways and/or familiar, and thus prosocial behavior ensues (see Brown, 2000 for a review of social identity theory). Similarly, this flexible distinction allows recategorization of group membership based on past behavior and the changing goals of the group and the individual. I disagree with this formulation because the metaphor for the in-group as the self assumes that the perception described above exists, which I doubt. Here I raise a question: What happens if the self and the in-group are considered as separate perceptual entities?

Evolutionarily, considering the self and the in-group as separate is not far fetched. Very few animals can recognize themselves, bringing into question whether a self-concept exists only in humans. Great apes, dolphins, and elephants (among other species) possess von Economo neurons,[2] pyramidal brain cells believed to be important in interoception or perception of bodily states (Allman et al., 2011; Stimpson et al., 2011). Self-perception ability is often tested with the mirror mark test (Amsterdam, 1972; Gallup, 1970), which involves placing the animal in front of a mirror with a mark, or some removable object, visible somewhere on the animal's body or face; if the animal is self-aware, it will attempt to remove the mark or object, or simply inspect it. This type of evidence is still somewhat controversial, and debate currently ensues. However, there is no debate that these animals do engage in group behavior. This suggests conceptions of the in-group may be separate from conceptions of the self, or alternatively, it is simply harder to test self-perception than group behavior. Assuming the former, one may find convergence in social psychological theory, where a self-concept is viewed as burdensome (Leary, 2004), and philosophy, where many doubt the self is substantial (Hume, 1739/1978) or can be defined (Locke, 1689/1975). Again, no such problems arise for conceptions of social groups. This suggests that group distinctions are evolutionarily older than self distinctions, and the group concept may have been a primary driving force during initial social interaction.

Then why the need for a self-concept? Most social interactions are filled with concerns about the self (Baumeister, 1982). In fact, most interpersonal interactions rely on comparing the relative position of the self in different hierarchies (Swencionis & Fiske, 2014). Therefore, one simple argument is that the self serves as a ready point of reference in social behavior. Indeed, another chapter can be written about self motives. But still, the organism is largely protected when belonging to a group, satisfying a hierarchy of needs (Maslow, 1943), allowing focus to be paid to other needs such as relative position within a group. But in migrant species such as humans, simply joining a group is a difficult task because people often have competing self and group motives. I attempt to make this clearer with the thought experiment at the beginning of this chapter that further describes how deception may have given rise to flexible social cognition.

Thought Experiment

On reviewing the thought experiment at the beginning of the chapter, some readers may think that killing by thoughts is rather far-fetched. However, people in groups are often motivated by ideology: a collection of thoughts. For instance, a member of the terrorist group Islamic State may die because of a thought: their ideology advocates suicide if it furthers their holy mission. Similarly, a member of the American military may also die from a thought: the patriotic belief in freedom may cause that person to place herself in dangerous situations to protect such freedom and her country, resulting in death. In fact, humans have been killed by thoughts, and have been motivated to kill other people because of thoughts, for much of our history. Additionally, throughout human history, people did live in small groups with their families and relatives, and a few non-kin (Thomas, 2006), so our example is perhaps not really too far-fetched.

In our thought experiment the governmental policy encourages migration between groups, allowing for a blending of the gene pool, so strangers are accepted into groups, as has occurred in human history. On encountering a stranger, how should a group member decide if (s)he is a good future group member or not (will not infect the group)? And for the stranger, suffice it to say that group membership brings benefits, such as an increased likelihood of survival and the opportunity to pass on one's genes. Nongroup membership certainly reduces the likelihood of survival and opportunities for procreation. Therefore, people are highly motivated to join groups whether they are already infected or not.

However, admitting someone to a group may result in harm since the increased cost to the group may not be made up by the positives the person adds. In actual human history a group member may have made physical evaluations of a stranger to begin addressing whether the increased costs of admitting this stranger compared to the benefits without considering outgroups, mental states, or intentions. Dissociating a potentially good future group member from a potentially bad future group member need not depend on mental state inferences. Therefore, whether humans actually used mental state inferences throughout their history for this task is debatable. However, in the thought experiment, such inferences are crucial. To the extent that the social group functions like a biological organism, which in a large sense it is, it has to make a decision about what it takes in and what it keeps out. People make those decisions regarding their body and food, lungs make them regarding air particles, and the brain makes them regarding chemicals. To help, human beings may rely on flexible social cognition to filter out deception.

Deception

But why might deception have promoted flexible social cognition? First, we must consider the nature of deception and what it means to pretend. If someone pretends, say a stranger pretends not to be infected, that person is communicating erroneous information. Since that information, such as desires, wishes, wants, and the like, does not exist beyond the private mind of the individual, there is no way of using external measures observable to everyone to determine whether the information (the person has the deadly thought, meaning that she is infected) actually occurs in the person's mind. Moreover, the fact is not *epiphenomenal* or just an explanation or reason if it actually drives the person's behavior (e.g., she will die from the virus, suggesting that she is capable of infecting others). For instance, an infected stranger may hold both the desire to help the group by being a good group member but may be well aware of the immense threat she poses for infection. That same desire to help may cause her to tacitly warn her new group members, infecting them in the process. Therefore, in her mind, the desire to help is primary, but it can drive harmful behavior.

Social cognition is critical to survival in this context. But not all forms of social cognition are suitable. For instance, imagining what the person feels, and feeling that way—engaging empathy—does not inform fully about whether that person is infected; it informs about the person's current emotional state, which is malleable. Here, I argue that flexible social cognition allows detection of deception because it is a spontaneous, brief appraisal process. This type of appraisal or inference entails assessments of the person's intentions and can result in the categorization harmer or helper, which is crucial to survival in this context. At this point the reader may argue that this appraisal that results an intentionality inference is the result of social cognition, not *flexible* social cognition. But I urge such readers to remember that in chapter 1, I described the human–nonhuman appraisal process as the essence of flexible social cognition. Therefore, flexible social cognition is necessary.

Intention

I argue that intentions are part of flexible social cognition as an initial appraisal process, but how does one go about inferring intentions, and what are intentions anyway? A perceiver decides that a target has good or ill intentions based on complex calculations that require the integration of statistical information (Heider, 1958; Gopnik & Wellman, 1992). This information includes the social context and the target as well as the person's own motivations and goals. For instance, consider a person walking down a dark alley. A stranger approaches her. If the stranger is very small, she may assume the stranger does not have bad intent. Similarly, if the alley is well lit, or very crowded, or on a Hollywood set, the stranger might not be perceived to have bad intent regardless of that person's size. Finally, if the perceiver is paranoid and delusional, or had just been mugged, then the stranger might be perceived to have bad intent regardless of size. These examples illustrate the impact of the social context on inferences of intent (as I discuss in detail in the next chapter) and highlight the fact that the agent alone does not provide sufficient information to infer intentionality. Therefore, social cognition inherently incorporates the agent, the social context, and the perceiver's own biases, personality, and experiences.

As discussed earlier, this ability to infer intentions is not reserved for human beings as targets and is displayed in other species as well. As the saying goes, even a dog knows when it has been kicked intentionally versus accidentally. This basic ability, however, may have given rise to the rich mental life most people experience and to other more evolved forms of social cognition as group size expanded beyond simply kin and a few close others.

Returning to our imagination, now let us consider a time in human history when groups are bigger. Human beings may still use the human–nonhuman appraisal evolved for small groups derived from flexible social cognition, which could enable more evolved, uniquely human social cognition. On encountering a stranger, the question about group status remains unknown. Physical indicators about the stranger may indicate group status. However, if these indicators are not available or the stranger is trying to deceive, then an inference of intent is the next best thing. What is different when we consider larger versus smaller social groups is that there exist in-group members who are also strangers.

Complex Mental Life

It is now possible for us to create a simple toy model describing the formation of complex mental life. Initial models for most species result in the following: core motives of survival and reproduction promote social interactions within and between social groups, but given deception, they require inferences of intentions (social cognition), or

S*R leads to sI leads to sC,

where (S*R) represents survival and reproduction motives, (sI) represents social interactions, and (sC) represents social cognition. However, for highly social species, difficult social contexts, such as scarce resources, impact the model. Survival and reproduction motives are now relatively heightened, changing the value of S*R. Obviously, events like murder continue to exist, suggesting that social cognition and even the resulting complex mental life are not sufficient to avoid death by social group, but they do seem necessary. Perhaps more primal motives often get in the way, or perhaps there is another intrinsic risk for death by complex mental life. Nonetheless, complex mental life did seem to result from social group formation and flexible social cognition.

The social intelligence hypothesis states that humans' superior intellect resulted from having to solve social problems. Because of their highly social nature humans faced unique problems surrounding social interaction. People with more complex cognitions and more intelligence better solved these problems because such abilities provided them a fitness advantage (Humphrey, 1976). This theory has received widespread support in the evolutionary anthropology literature, but it features less prominently in other fields. Nonetheless, it suggests that our social context promoted our increased cognitive abilities, including complex mental life and other forms of meta-cognition. Since social cognition may have also evolved from similar social pressures, it suggests that social cognition represents a form of intelligence, or complex cognition at a minimum.

However, given that other agents infer intentions as well, how could social cognition, or rather flexible social cognition more specifically represent uniquely human complex cognition? There is evidence that dogs possess social cognitive abilities (Agnetta, Hare, & Tomasello, 2000); dogs use pointing cues to find food, suggesting that they share joint attention abilities (a form of social cognition) with humans (as discussed in chapter

3). Therefore, social cognition may not be uniquely human, and it may be sufficient but not necessary to infer intention: it may be possible to infer intention without social cognition. An inference of intentionality results from social cognition, but so do other things; moral protection, reserved for human beings, and subsequent empathic processing also result from social cognition. There is currently debate whether these abilities are also unique to humans. Some researchers have claimed that bacteria are prosocial and altruistic (Youk & van Oudenaarden, 2010); a few *Escherichia coli* cells will sacrifice themselves in the presence of antibiotics, causing their own death but ensuring survival of the population of cells. Therefore, other species may be able to infer intentionality, and may be able to display prosocial behavior but may lack the ability to engage social cognition. Even if they do have the capacity for social cognition, they may lack the ability for flexible social cognition, which may be unique to human beings.

Perhaps social interaction may have advantaged social cognition, but that does not make it a form of complex cognition. Stated differently, not all processes favored by evolution for their social advantage are necessarily complex mental constructs. This may indeed be true of social cognition. Nonetheless, this particular debate addresses concerns relative to whether all agents that display social cognitive abilities display complex cognition. Moral arguments often hinge on the separability of human and nonhuman psychological experiences, determining whether such nonhuman agents deserve to be treated more like humans, avoiding captive existences, brutality, and slaughter for food or sport. Sadly, humans themselves still suffer from these behaviors, making the plea for nonhuman animals all the more distant.

Thus far, I have described flexible social cognition as quick, spontaneous, and beyond conscious control. These features describe an automatic, heuristic, system 1 process, not a complex cognitive construct that typically resides in the more controlled system 2.[3] If social cognition does indeed reflect complex cognition, then it suggests that the definitions of automatic and controlled processes need revision. Specifically, there may be forms of complex cognition that function spontaneously, quickly, without conscious awareness. Given that there are four criteria that need to be met to describe a process as automatic (Bargh, 1994), few phenomena meet these criteria. Social cognition also falls short. I next consider the impact of the social context on social cognition.

Notes

1. There is some highly disputed evidence that nonhuman animals and even plants can deceive. For instance, many plants evolve to appear as not harmful to their prey, such as the well-known Venus flytrap (*Diomaea muscipula*). When insects touch hairs on the surface of a leaf, the closing mechanism of the leaf is triggered, trapping the insect. It is debatable whether this example depicts deception or simply some form of trap, like a mousetrap. Other evidence comes from animals that camouflage themselves, hiding from predators, such as some species of toads, owls, caterpillars, and spiders. Although these animals are clearly agents, it is difficult to infer deception based on their behavior. Most convincing perhaps is evidence of deception among animals in competitive contexts. For instance, rhesus macaque monkeys (*Macaca mulatta*), when given a choice between an experimenter with a food reward who is facing the monkey and another experimenter with a food reward whose back is turned, will take food from the latter experimenter (see Drayton, Varman, & Santos, 2016). All these examples hint at deception but do not provide strong evidence.

2. Von Economo neurons (VENs) are spindle neurons, a type of pyramidal cell, found in parts of the neocortex, including ACC, insula, and DLFPC. These cells are structurally different from other pyramidal cells because of their single axon with a single facing dendrite. Such cells are a rare brain cell type and are only found in a handful of creatures, including humans. However, these cells are abundant in humans, and they are often thought of as partially responsible for human uniqueness.

3. Both social and cognitive psychology have described dual-process models of cognition that distinguish between automatic and controlled processes (see Kahneman, 2011). System 1 processes, or automatic processes, are those that occur without awareness, unintentionally, without control, and effortlessly, whereas system 2, controlled processes require awareness, intention, control, and are effortful. For instance, learning to drive a car is an effortful, system 2 process. However, once the skill has been acquired and a driver has enough experience, driving can become a system 1, automatic process. But it turns out that there are very few processes, if any, that meet all the requirements to be considered system 1. For instance, an experienced driver is still in control of driving. More recently, researchers in the fields of social and cognitive psychology have whispered among themselves that the distinction between system 1 and system 2 is a false dichotomy, and most processes are somewhere in the middle. This harkens back to the single-mode alternative theory first extolled during the height of the automatic and controlled zeitgeist that received little attention at the time (see Kunda, 1999; Kunda & Thagard, 1996). This theory argued that all types of processes are equivalent. For instance, in impression formation, dual-process models argue that system 1 information (such as category membership that triggers stereotypes) influences the final impression, particularly if the person is not motivated to form an impression. The single-mode alternative argues that all types of information are used, regardless of the person's motivation.

References

Agnetta, B., Hare, B., & Tomasello, M. (2000). Cues to food location that domestic dogs (*Canis familiaris*) of different ages do and do not use. *Animal Cognition*, 3(2), 107–112.

Allman, J. M., Tetreault, N. A., Hakeem, A. Y., Manaye, K. F., Semendeferi, K., Erwin, J. M., et al. (2011). The von Economo neurons in the frontoinsular and anterior cingulate cortex. *Annals of the New York Academy of Sciences*, 1225(1), 59–71.

Amsterdam, B. (1972). Mirror self-image reactions before age two. *Developmental Psychobiology*, 5, 297–305.

Bargh, J. A. (1994). The four horsemen of automaticity: Awareness, intention, efficiency, and control in social cognition. In R. Wyer & T. Srull (Eds.), *Handbook of social cognition,* 1 (pp. 1–40). Mahwah, NJ: Lawrence Erlbaum.

Baumeister, R. F. (1982). A self-presentational view of social phenomena. *Psychological Bulletin*, 91(1), 3–26. doi:10.1037/0033-2909.91.1.3.

Brown, R. (2000). Social identity theory: Past achievements, current problems and future challenges. *European Journal of Social Psychology*, 30(6), 745–778.

Dovidio, J. F., Gaertner, S. L., & Saguy, T. (2008). Another view of "we": Majority and minority group perspectives on a common ingroup identity. *European Review of Social Psychology*, 18(1), 296–330. doi:10.1080/10463280701726132.

Drayton, L. A., Varman, L., & Santos, L. R. (2016). Capuchins (*Cebus apella*) are limited in their ability to infer others' goals based on context. *Journal of Comparative Psychology*, 130, 71–75.

Fiske, A. P. (1992). The four elementary forms of sociality: Framework for a unified theory of social relations. *Psychological Review*, 99(4), 689.

Gallup, G. G. (1970). Chimpanzees: Self-recognition. *Science*, 167, 86–87.

Gopnik, A., & Wellman, H. M. (1992). Why the child's theory of mind really is a theory. *Mind & Language*, 7(1-2), 145–171.

Heider, F. (1958). *The psychology of interpersonal relations* Hoboken, NJ: John Wiley & Sons. 10.1037/10628-004.

Hume, D. (1739–1740/1978). *A treatise of human nature*, L. A. Selby-Bigge (Ed.), revised by P. H. Nidditch. Oxford: Oxford University Press.

Humphrey, N. (1976). The social function of intellect. In P. P. G. Bateson & R. A. Hinde (Eds.), *Growing points in ethology* (pp. 303–317). Cambridge, UK: Cambridge University Press.

Kahneman, D. (2011). *Thinking, fast and slow*. New York: Farrar, Straus and Giroux.

Kunda, Z. (1999). *Social cognition: Making sense of people*. Cambridge, MA: MIT Press.

Kunda, Z., & Thagard, P. (1996). Forming impressions from stereotypes, traits, and behaviors: A parallel-constraint-satisfaction theory. *Psychological Review*, 103(2), 284.

Leary, M. R. (2004). *The curse of the self: Self-awareness, egotism, and the quality of human life*. New York: Oxford University Press.

Locke, J. (1689/1975). *An essay concerning human understanding,* P. H. Nidditch (Ed.). Oxford: Oxford University Press.

Maslow, A. H. (1943). A theory of human motivation. *Psychological Review*, 50(4), 370–396.

Stimpson, C. D., Tetreault, N. A., Allman, J. M., Jacobs, B., Butti, C., Hof, P. R., et al. (2011). Biochemical specificity of von Economo neurons in hominoids. *American Journal of Human Biology*, 23(1), 22–28.

Swencionis, J. K., & Fiske, S. T. (2014). How social neuroscience can inform theories of social comparison. *Neuropsychologia*, 56, 140–146. doi:10.1016/ j.neuropsychologia.2014.01.009.

Thomas, E. M. (2006). *The old way: A story of the first people*. New York: Macmillan.

Youk, H., & van Oudenaarden, A. (2010). Altruistic defence. *Nature*, 467, 34.

7 The Social Context

It is clear that we cannot distinguish the sane from the insane in psychiatric hospitals. The hospital itself imposes a *special environment* in which the meanings of behavior can easily be misunderstood. The consequences to patients hospitalized in such an environment—the powerlessness, depersonalization, segregation, mortification, and self-labeling—seem undoubtedly countertherapeutic.

—D. L. Rosenhan (1973), italics not included in original.

Social psychological research documents the power of the social context: the excerpt above describes one such contextual impact, where researchers were institutionalized simply because they showed up at a psychiatric facility and claimed to hear voices. Subsequent "sane" behavior was considered additional evidence of illness, and they faced great difficulty convincing the administrators they were not insane.

An underlying truth about humans is that the social context determines reality. Meta-analyses suggest over 50 million participants are influenced by the social context in scientific experiments (Fiske, Harris, & Cuddy, 2004). But what is meant by social context? And how has it come to exert such influence? The social context refers to the physical and social environment in which behavior occurs. For instance, imagine having ice cream on the beach in summer versus winter, or going out for ice cream with a warm or cold person. Each describes a physical and social environment, and the alternative in each scenario is rather unpleasant, perhaps changing your behavior toward the ice cream and the person as well.

The social context can affect not only brain and physiological responses, driving behavior, but the likelihood of behavior being performed in the future by other people. This is because the social context creates culture, transmitting behavior as moral rules and norms vicariously from one person to another in a society. Research in epigenetics suggests that ribonucleic acid (RNA) messenger molecules in honey bees and behavior changes (such as transitioning from one work role to another) are coupled (Whitfield, Cziko, & Robinson, 2003), resulting in changes in the proteins that constitute the characteristics of the cells (including brain cells). Because RNA is impacted by the environment, the underlying biological mechanism that originates behavior is susceptible to the social context (for an interesting discussion on the role of other contextual factors such as bacteria on RNA and brain evolutions, see Stilling, Dinan, & Cryan, 2014). This more recent finding nullifies centuries old debates about nature versus nurture, suggesting that each impacts the other; underlying inherited biology is shaped by the context, and behavior in the context results from the underlying biology. This fundamental feature of contexts suggests that they are powerful tools for shaping behavior. Moreover, understanding the interaction of context with biology creates a more comprehensive picture of the mechanisms responsible for social behavior.

Before we consider the social context in more detail, it is important to note that it is not the only motive driving behavior. Factors internal to the person, such as her genetics, traits, tendencies, experiences, hormones, and physical and mental capabilities also help to determine behavior. Much ink has been spilled, and keys tapped, describing how these often idiosyncratic, individual difference factors drive behavior. The social context interacts with these factors to determine behavior. My focus on the social context is not to diminish the importance of individual difference factors but, rather, to use the social context to tell a story about what may have happened on average to human beings. In fact, a feature of the social context that makes it extremely influential is that it provides a baseline for behavior from which variance driven by individual differences may occur.

Let us now return to our thought experiment from the previous chapter. Imagine encountering a stranger in a more evolved human social context where societies have replaced small groups. Importantly, competition between groups now exists and can at times be hostile and intense. How do you engage a stranger? As a potential ally or foe? Here, revisiting animal behavior proves useful. As discussed in chapter 2, there are a few closely related species who differ in their death by social group behavior toward strangers. Chimpanzees, evolving north of the Congo River, engage in hostile, xenophobic behaviors toward strangers found during pack territory patrol (Fawcett & Muhumuza, 2000). This death by social group is also found in wolves (Scott, 1997). However, murder is absent in the closely related bonobo and dog behavior respectively. Bonobos are primarily thought to have evolved south of the Congo River. There is less food competition from other apes south of the river, making a variety of food abundant. North of the river, chimpanzees face competition for food from gorillas and other chimpanzees. Bonobos instead engage in xenophilic behaviors, practicing nonreproductive sexual behavior with strangers, not murdering them. Scientific experiments capture differences in willingness to share food between bonobos and chimpanzees (Feistner & McGrew, 1989; Tan & Hare, 2013), although both display inequity aversion (Brosnan, Schiff, & De Waal, 2005).

Experiments have also demonstrated that foxes that are more docile and friendly toward humans become more dog-like over successive generations if this gene is promoted during breeding (Trut, Oskina, & Kharlamova 2009). Although one may argue for the influence of genetics, an individual difference factor, the social context has been created by the minx fox farmers who provided breeding opportunities to the foxes, tilting the influence of genetics. This example demonstrates the interaction of social context and individual differences.

Therefore, the social context matters. A harsh context with intergroup hostility and limited resources makes determining the mental states of a stranger crucial. Engaging in murder may deny the in-group a valuable new member; yet admitting the member and trusting her may put the group at risk. Because of deception, a simple mental state inference the moment one first encounters a stranger may not be sufficient. Why? Because the inference is fleeting and does not occur beyond the mind of the perceiver. Stated differently, the inference was not a perception, processed bottom-up through the retina or some other sensory organ, relayed to the primary sensory cortex then onto the remainder of the brain to be stored as memory, and so on. The inference may have been triggered by an external stimulus (the stranger) but is a logical conclusion about that stimulus and its behavior, not a perception of the stimulus or behavior itself. Because of the power of the social context, the same behavior from the same stranger could lead to a different mental state inference. Therefore, the inference reflects the interaction of the external stimulus with the social context and is unique to that particular moment in time. Fundamentally, this is how the social context moderates social cognition. But this opinion suggests social cognition engagement is insufficient for preparing action tendencies toward other people. Indeed in our example, the mental state inference may have been driven more by the stranger or the social context, so the inference itself is insufficient. More inferences may be needed, even many inferences over time, or knowledge about the impact of the social context on other people. In other words, a complex mental life may be needed—a construct capable of supporting multiple inferences about people with many baselines for behavior represented by the social context.

Classic Social Psychology

Beginning with Kurt Lewin (1939/2009), social psychology stakes a claim as the academic discipline that empirically studies the social context. This may seem a bit strange since one might believe that a social psychologist should study people, that is, individual differences. Yet that task is often left for personality psychologists. Since

$$M_t = f(a, s)$$

personality psychology focuses on *a* and social psychology *s*. Although both fields do explore the interaction, social psychology has, throughout its academic history, shined a light on the often ignored social context. In fact, some of the most famous experiments in the field highlight the power of the social context.

Take for instance the bystander intervention effects. Inspired by the grisly murder of Kitty Genovese in New York City in 1964, social psychologists wondered whether the neighbors who watched the murder from their apartments[1] did not act because of their cold, callous nature (an explanation of the behavior focusing on *a*) or because the social context did not fully diagnose the situation as one that required them to take action and help (an explanation of the behavior focusing on *s*). While the press made the fundamental attribution error—focusing on the inferred disposition and inherent morality of New Yorkers rather than the social context—social psychologists demonstrated that such situations lead to a diffusion of responsibility: each person thinks the other is going to help. The more people not taking action, the less the situation seems like one that requires action.

To demonstrate this effect, Darley and Batson (1973) ran an experiment that elicited stranger apathy. Participants were recruited from a theological seminary. They were instructed that they must prepare a sermon on the Good Samaritan parable—a story that described helping and praises the practice. But participants were told that they had to deliver that sermon very soon across campus from the research lab where they were receiving these instructions. Therefore, they were urged to hurry over to that location. In the alley next to the lab on the way across campus lay a confederate (fake participant planted by the experimenters to elicit behavior from the participant), hunched over. The researchers measured how many people, thinking about the Good Samaritan parable, would stop to help the confederate. Very few did, suggesting that even people presumable thinking about helping behavior did not diagnose a situation as such because of their rush to make an appointment (there are also other interpretations of this result, but for the current purpose, I will stick with the classic social psychological interpretation). In another such experiment (Darley & Latane, 1968), participants were speaking with a confederate on an intercom when suddenly the confederate faked a seizure. The manipulated variable was how many other people were in the room with the participant. If a confederate was present and did not respond to the emergency, the participant was less likely to respond. This effect has been replicated with smoke coming into the participant room instead of a confederate needing help as the emergency, with the novel situation posing a threat to the participants themselves (Latane & Darley, 1968). Yet still, if a confederate in the room remained calm and did not take action, a participant was more likely to do the same. This illustrates the power of the social context: even when the participants' lives were potentially in danger, they did not take action to save themselves because of the information gleaned from the social context.

Social psychology has demonstrated similar impacts of the social context in harming scenarios as well. Inspired by the Holocaust and Hannah Arendt's (1963) description of Nazi perpetrators as motivated by the social context and blindly following authority, social psychologists showed that obedience to authority is another situational factor that can drive antisocial behavior. Stanley Milgram's (1963) obedience study led participants to believe that they could administer electric shocks to a confederate based on that person's performance. The experimenter, in authoritative white lab coat, stood by and simply requested that the participant increase the voltage and continue to shock the confederate, who now yelled in pain as the shocks increased. It should be noted that the electric shocks were not really administered, and the confederate faked the entire shock experience. In most but not all circumstances, most people continued to comply with the experimenter, even when the confederate stopped responding to the shocks, suggesting something drastic had occurred.

Similar examples of obedience to authority may be found in cases of torture during hostile conflicts. American and British soldiers have tortured prisoners of war in Iraq, Afghanistan, and Guantanamo Bay. These are not isolated incidents of a few bad apples but rather reflect the power of the social context in creating an environment where prisoner torture was acceptable. Constantly having your life under threat, being shot at, being far from home in unfamiliar circumstances, with a chain of command that impresses on you the importance of extracting information from prisoners, dehumanizing and calling prisoners derogatory names—these factors all combine to create a social context suitable for prisoner torture (Fiske, Harris, & Cuddy, 2004).

Of course, examples of prisoner abuse harken to the most famous empirical case: an actual social psychology study of prisoner and guard behavior. In this study Philip Zimbardo (1973) randomly assigned participants to either the role of prisoner or prison guard. The researchers made the social context as realistic as possible including having participants wear the appropriate clothing and commandeering the basement of a psychology department to serve as holding cells. Over the course of two weeks, participants slowly conformed to the social roles, where prisoners verbally abused guards, and guards verbally and physically abused prisoners. In fact, the experimentally created social context was so powerful that the experiment had to be prematurely ended for participant safety. This experiment demonstrates that even when people are inherently aware of the fictitious nature of the social context, it still exerts an effect on their behavior and social cognition.

Perhaps the most compelling evidence demonstrating the power of the social context came when social psychologists pretended to be insane. In this experiment by David Rosenhan (1973), confederates served as "pseudo" patients, reporting to a mental health facility claiming that they heard voices. All of the confederates were easily admitted to the facility. Remarkably, they faced immense difficulty when attempting to leave the facility; no one in the administration believed they were not actually insane (or else why would they have committed themselves)? This example beautifully illustrates that, despite arguments to the contrary, the social context creates expectations about other people in that context.

Types of Social Contexts

So what types of social contexts exist in modern human society. Surely, the small-group context no longer exists as a singular group; instead, people now belong to multiple small and larger groups based on multiple social identities. A group becomes relevant when an identity is made salient in a social context (Tajfel & Turner, 1979). Building on the classic social identity framework, these multiple groups can be categorized perhaps as follows: altruistic, economic, legal, family, close friends, office/formal work, religious, temporary, friendly competitive, and antagonistic competitive. Now this listing may seem arbitrary (and to some degree it is), but it provides a basis to describe different dimensions that govern social behavior; each of these dimensions interacts to produce each of the above contexts. Moreover, these categories are not mutually exclusive, and often overlap, suggesting that multiple dimensional interactions can themselves interact (and most statisticians would probably struggle to clearly explain interactions with three or more main effects because of the amount of simultaneous comparisons). Therefore, the behavior of each individual in a unique social context at a particular moment in time is multiply determined by a number of contextual factors in addition to the contribution of individual differences.

Researchers in anthropology have taken this descriptive approach to human societies, labeling them as communal, reciprocal, market exchange, and barter and trade (Fiske, 1992; McGraw & Tetlock, 2005; McGraw, Tetlock, & Kristel, 2003). Such broad categories may apply to groups as well, and they may describe the dimensions along which behavior may be determined. However, because of the complex interactions between and among dimensions just described, it may be best to simply consider the types of multiple groups individually, describing these before considering unique interactions.

Restoring Some Autonomy: Consistency Theories

To this point, the discussion of the social context may suggest that people are akin to automatons and guided by situational factors rather than intentions and volition. However, motives drive people's behavior in each social context. Such motives can be large and abstract, or minor and concrete. We find this illustrated by *implementation intentions*: making an if-then statement about intended behavior is an example of a smaller motive for behavior because it is constrained to a very narrow social context and behavior (Gollwitzer, 1999). By contrast, larger, more abstract motives such as consistency theories—motives regarding keeping behavior and attitudes consistent—govern across many more social contexts.

Social psychological "balance" theories all describe motivational processes that seek to reduce inconsistencies between psychology and behavior. For example, balance theory describes a sequence of relationships among three individuals (or two and an object; Heider, 1946). It states that there must be a harmony among the valence of the relationships, such that there is mutual dislike among enemies, or liking among friends. Therefore, if person A likes person B, and person B likes person C, then person C must like person A for there to be balance within these relationships. However, if person C does not like person A, then this causes imbalance, and the warmth or liking in the relationship between person A and person B, and between person B and person C is threatened. Similarly, if person A dislikes person B, and person B dislikes person C, then person C must dislike person A for there to be balance within these relationships. However, if person C likes person A, then this causes imbalance, and the coldness in the relationship between person A and person B, and between person B and person C is threatened (this theory often nicely explains the conflict when a close family member forms a close relationship with a disliked other, be it out-group member, or simply a person not liked by the initial family member). The theory also works if an object or activity that is either liked or disliked replaces person C.

Other social psychological theories also emphasize consistency. Cognitive dissonance theory argues that people are motivated to have consistency between their attitudes and behavior. For instance, people receiving insufficient reward for dishonest behavior will come to believe that they do hold the dishonest opinion (Festinger & Carlsmith, 1959). In this classic study participants were first made to complete a very boring task; they were instructed to turn a large number of pegs in a wall a small degree, and repeat the action a number of times. Participants were then told that the experiment had ended, but unfortunately, the research assistant who helped out with participants was absent that day. As a result, the experimenter needed someone to tell the next participant about the task, and given that the participant had just completed the task, he would be an excellent source of information for the new participant. However, it was important that he make the task appear interesting to the new participant. The experimenter then either gave the person a relatively large sum of money ($20, sufficient justification) or a relatively small sum of money ($1, insufficient justification) to lie to the new participant about the true nature of the task. Most participants complied, telling the new participant that the task was interesting and engaging. On leaving the experiment lab, the participants then encountered another person who was collecting information about participants' experiences when completing psychology experiments. This person then asked the participant to report on how engaging and interesting they had found their recently completed experimental task. Participants in the insufficient justification condition were more likely to report that they had found the task very engaging and interesting than participants with sufficient justification. These results suggest that the participants who had only been paid $1 to lie to someone else, condemning that new participant to a negative outcome (since they now had to complete a very boring task), changed their opinion about the task to be in line with their behavior; that is, they now had convinced themselves that the task actually had been interesting.

Another demonstration of this consistency active during cognitive dissonance comes from a slightly different experimental paradigm. Participants were shown a range of household products, and asked to rate them according to how desirable they found each product. The experimenter then presented the participant with a choice between two of the products and informed the participants that the chosen product would be theirs to take home at the end of the experiment. However, the experimenter manipulated the choice between subjects; some participants were shown two products that differed in their ratings (one was rated as highly desirable, the other not), whereas some participants were shown two products that were rated similarly on desirability. After participants chose a product, the experimenter then had them rerate the entire set of products. Participants who had to choose between similarly rated products then devalued the unchosen option, and thereby increased the rating of the chosen option, while there was no change for participants given the choice between initially dissimilar products (Brehm, 1956). Presumably, this occurred to justify the choice between two similar products such that the unchosen option was never really very much liked, whereas the chosen option was always liked a lot.

Further social psychological evidence demonstrates consistency in another way; people will come to believe that others hold particular opinions after hearing them expressed even when told that the expression did not reflect the author's true opinion (Jones & Harris, 1967). In this study, which took place in America when tensions between the United States and Cuba were high, participants were asked to listen or read either a pro-Castro, anti-Castro, or equivocal (containing pro- and anti-Castro sentiments) essay. Participants were instructed that despite the author's true beliefs, writing the pro-Castro essay was necessary because more positive opinions were needed to represent an alternative point of view, given that almost everyone wanted to write an anti-Castro essay. Participants then reported the attitudes they perceived the author of the essay held about Fidel Castro. Remarkably, participants who had listened to or read a pro-Castro essay believed the author held more positive attitudes of the Cuban leader. Again, the researchers assumed this occurred because participants needed to achieve consistency between the perceived attitudes of the author toward Castro and their behavior (writing a pro-Castro essay). Importantly, the participants in this study were told that the author was not forced to write a pro-Castro essay, but were gently encouraged; strong coercion would have provided justification for the behavior.

People are also more likely to buy a larger product or make a larger commitment after having already made a smaller commitment (Freedman & Fraser, 1966). In one example of this effect, participants were asked to answer questions about the kinds of soaps they used. Following compliance with this relatively small request, participants were much more likely to allow a survey team to enter their homes for two hours to catalogue their household products. Similarly, participants who were asked to sign a petition or put a small sign on their cars advocating a political issue were more likely to consent to placing a large sign with the political message on their front lawn. This foot-in-the-door effect also hinges on a motive for consistency; if the participant consented to the smaller request, then consenting to the larger request is simply consistent with her earlier behavior.

People also strive for consistency in their sense of self. Self-perception theory argues that people view themselves as consistent with their behaviors (Bem, 1972). For instance, a person who is made to wear a lab coat will often view himself as much more intelligent and be willing to attempt to solve more complex problems. Related to self-perception theory, stereotype threat theory suggests that people perform in line with stereotypes if made aware of those stereotypes before performance (Steele & Aronson, 1995). Therefore, Asian women who were about to take a math test could either perform well, consistent with an Asian stereotype suggesting that such ethnic group members are good at math, or perform poorly consistent with a female stereotype that women are bad at math. In such an experiment the experimenter asked Asian female participants to indicate either their gender or their ethnicity before completing a math test. Participants who indicated their gender performed more poorly than participants who indicated their ethnicity. This suggests that the activated stereotype led participants to behave in a consistent manner (e.g., Sinclair, Hardin, & Lowery, 2006) and respond to a "stereotype threat in the air" according to the researchers (Steele, 1997).

People even refuse to change stereotypes, preferring to subtype or to view violations as outliers (see Neugarten, 1974, for early example). This occurs in order to maintain a consistent stereotype about the group. For example, a person who stereotypes people of African descent as unintelligent, yet encounters a person of African descent who is intelligent, may subtype that person; that is, she may create a new category to ascribe that person, perhaps a category of smart people of African descent, rather than simply a person of African descent. This preserves the negative stereotype while it accounts for the stereotype-inconsistent group member. Alternatively people may perceive the inconsistent group member as an outlier, not a true representation of the group. This also achieves consistency of cognitions because it excludes that specific inconsistent group member from the group, preserving the negative group stereotype.

This raft of classic social psychology research just discussed demonstrating respectively cognitive dissonance (insufficient justification and forced-choice paradigms), the "Castro effect," the foot-in-the-door effect, self-perception, stereotype threat, and subtyping suggests that motives drive human beings to achieve consistency. Mental construct tension functions on the same fundamental human mechanisms—interacting with the situation to create behavior.

A Final Word on the Social Context

I often remember listening to faculty in graduate school describe the desire to specify a periodic table of human behavior. The argument proceeded that if chemistry has such a table of elements, and psychology is indeed a science, and behavior is a natural kind just like biological or physical properties of objects, then human behavior should be reducible to its elementary components, and psychologists should be able to create such a table illustrating the relationships between/among these components. Of course, such a challenge has not yet been met by psychology, providing ammunition for detractors who question the validity of the scientific enterprise. Perhaps like any good null result, the solution may yet be found in the future, and such a periodic table of behavior can be designed. Or perhaps the comparison to chemistry is the wrong one, and instead of a periodic table of behavior, a bandwidth of signals is a more apt analogy.

Specifically, perhaps the brain registers signals that communicate information about the social context. This then determines the appropriate behavior for a person given information about her social role within the context, as well as the person's biological limitations and propensity to display over-learned behavior. This approach identifies four components that may determine behavior: (1) social contextual signal, (2) social role/identity, (3) biology/traits, and (4) predispositions for learned behavior. Therefore, determining the bandwidth of social contextual signals would further specify behavior. This social contextual signal would tag all subsequent psychological processing but is itself the result of psychological processing. Since both the processes it tags and the ones that arise from the tagging are psychological processes, this signal then sits at the junction, and determining the various properties of this signal can cast further light on both the psychological and biological underpinnings of behavior. But I reserve that task for another book. Hoping that I have convinced you about the power of the social context, I will now describe two social contexts in more detail to illustrate the impact of flexible social cognition.

Note

1. There are many distorted reports of this incident regarding the extent to which neighbors could witness the incident, but it nonetheless motivated social psychological research.

References

Arendt, H. (1963). *Eichmann in Jerusalem*. New York: Penguin.

Bem, D. J. (1972). Self-perception theory. *Advances in Experimental Social Psychology*, 6, 1–62.

Brehm, J. W. (1956). Postdecision changes in the desirability of alternatives. *Journal of Abnormal and Social Psychology*, 52(3), 384–389.

Brosnan, S. F., Schiff, H. C., & De Waal, F. B. (2005). Tolerance for inequity may increase with social closeness in chimpanzees. *Proceedings of the Royal Society of London. Series B, Biological Sciences*, 272(1560), 253–258.

Darley, J. M., & Batson, C. D. (1973). "From Jerusalem to Jericho": A study of situational and dispositional variables in helping behavior. *Journal of Personality and Social Psychology*, 27(1), 100–108. doi:10.1037/h0034449.

Darley, J. M., & Latane, B. (1968). Bystander intervention in emergencies: Diffusion of responsibility. *Journal of Personality and Social Psychology*, 8(4), 377–383.

Fawcett, K., & Muhumuza, G. (2000). Death of a wild chimpanzee community member: Possible outcome of intense sexual competition. *American Journal of Primatology*, 51(4), 243–247.

Feistner, A., & McGrew, W. (1989). Food-sharing in primates: A critical review. In P. Seth & S. Seth (Eds.), *Perspectives in primate biology* (Vol. 3, pp. 21–36). New Delhi: Today & Tomorrow.

Festinger, L., & Carlsmith, J. M. (1959). Cognitive consequences of forced compliance. *Journal of Abnormal and Social Psychology*, 58(2), 203–210.

Fiske, A. P. (1992). The four elementary forms of sociality: Framework for a unified theory of social relations. *Psychological Review*, 99(4), 689–723.

Fiske, S. T., Harris, L. T., & Cuddy, A. J. C. (2004). Why ordinary people torture enemy prisoners. *Science*, 306(5701), 1482–1483.

Freedman, J. L., & Fraser, S. C. (1966). Compliance without pressure: The foot-in-the-door technique. *Journal of Personality and Social Psychology*, 4(2), 195–202.

Gollwitzer, P. M. (1999). Implementation intentions: Strong effects of simple plans. *American Psychologist*, 54, 493–503.

Heider, F. (1946). Attitudes and cognitive organization. *Journal of Psychology*, 21, 107–112.

Jones, E. E., & Harris, V. A. (1967). The attribution of attitudes. *Journal of Experimental Social Psychology*, 3(1), 1–24.

Latane, B., & Darley, J. D. (1968). Group inhibition of bystander intervention in emergencies. *Journal of Personality and Social Psychology*, 10(3), 215–221.

Lewin, K. (2009). Experiments in social space (1939). In B. F. Gentile, & B. O. Miller (Eds.). *Foundations of psychological thought: A history of psychology* (pp. 454–467). Thousand Oaks, CA: Sage Publications.

McGraw, A. P., & Tetlock, P. E. (2005). Taboo trade-offs, relational framing, and the acceptability of exchanges. *Journal of Consumer Psychology*, 15(1), 2–15.

McGraw, A. P., Tetlock, P. E., & Kristel, O. V. (2003). The limits of fungibility: Relational schemata and the value of things. *Journal of Consumer Research*, 30(2), 219–229.

Milgram, S. (1963). Behavioral study of obedience. *Journal of Abnormal and Social Psychology*, 67(4), 371.

Neugarten, B. L. (1974). Age groups in American society and the rise of the young-old. *Annals of the American Academy of Political and Social Science*, 415(1), 187–198.

Rosenhan, D. L. (1973). On being sane in insane places. *Science*, 179(4070), 250–258.

Scott, J. P. (1997). Aggression, violence, and culture. In D. Munro, J. F. Schumaker, S. C. Carr, D. Munro, J. F. Schumaker, & S. C. Carr (Eds.), *Motivation and culture* (pp. 224–233). New York: Routledge.

Sinclair, S., Hardin, C. D., & Lowery, B. S. (2006). Self-stereotyping in the context of multiple social identities. *Journal of Personality and Social Psychology*, 90(4), 529–542.

Steele, C. M. (1997). A threat in the air: How stereotypes shape intellectual identity and performance. *American Psychologist*, 52(6), 613–629.

Steele, C. M., & Aronson, J. (1995). Stereotype threat and the intellectual test performance of African Americans. *Journal of Personality and Social Psychology*, 69(5), 797–811.

Stilling, R. M., Dinan, T. G., & Cryan, J. F. (2014). Microbial genes, brain & behaviour—epigenetic regulation of the gut–brain axis. *Genes Brain & Behavior*, 13(1), 69–86.

Tan, J., & Hare, B. (2013). Bonobos share with strangers. *PLoS One*, 8(1), e51922.

Tajfel, H., & Turner, J. C. (1979). An integrative theory of intergroup conflict. In W. G. Austin & S. Worchel (Eds.), *The social psychology of intergroup relations* (pp. 33–47). Pacific Grove, CA: Brooks/Cole.

Trut, L., Oskina, I., & Kharlamova, A. (2009). Animal evolution during domestication: The domesticated fox as a model. *BioEssays*, 31(3), 349–360.

Whitfield, C. W., Cziko, A. M., & Robinson, G. E. (2003). Gene expression profiles in the brain predict behavior in individual honey bees. *Science*, 302(5643), 296–299.

Zimbardo, P. G. (1973). On the ethics of intervention in human psychological research: With special reference to the Stanford prison experiment. *Cognition*, 2(2), 243–256.

8 The Economic Context

A person walks into a casino. In this situation she is faced with two kinds of games with different probabilities of winning money: those versus machines (e.g., slot machines, roulette) and those versus other people (e.g., blackjack, poker). But does the person employ the same strategy when playing against these two different competitors? Yes, she may wonder whether the machine is broken, and therefore reliable, but does she distrust the machine the way she may distrust the shifty blackjack dealer? She decides to play blackjack—and promptly loses all her money.

People make different economic decisions regarding other people and objects (see Lee & Harris, 2013 for review). Given that other people have minds capable of forming an impression, people are constantly self-monitoring to ensure that said impression is favorable. People are also capable of deception; this agentic feature of people results in uncertainty during economic interactions. It is possible that a person may self-monitor, appearing trustworthy and reliable to others when she actually harbors ill will and is motivated strictly by self-interest, often at the cost of the other person in the economic exchange. Solving such problems requires additional cognitive processing. Neuroimaging research suggests that when people co-opt brain systems involved in economic decisions and subjective value such as the striatum and medial orbitofrontal cortex (MOFC) for social economic exchanges, they also rely on their social cognition brain network, presumably to provide additional cognitive processing muscle (Lee & Harris, 2013; Rilling & Sanfey, 2011). Therefore, people in social economic contexts are concerned with not only market forces but social forces as well.

Human beings display aversion to inequity (Adams, 1965). This trait runs counter-productive to hierarchical systems prevalent throughout the animal kingdom, including in human societies (although whether nonhuman primates also display inequity aversion is hotly debated; see Silberberg, Roma, Ruggerio, & Suomi, 2006). How is this conflict resolved? That is, how do people live in hierarchical structures yet despise inequity? Many structural forces quell the conflict, although within limits. For instance, *deservingness* is attached to high status and power, justifying inequality (see Bobocel, Kay, Zanna, & Olson, 2010, for discussion). Similarly, in American society, a Protestant work ethic "justifies" inequality since the perception exists that the system itself is fair and just (Rosenthal, Levy, & Moyer, 2011): anyone who works hard, it contends, can ascend the social hierarchy. Another example comes from religion, which has long been described as a panacea for the masses and a form of control over reactions to inequality (Raven, 1999). Economic systems also serve the same function by reducing the inconsistency between inequity aversion and hierarchical systems.

A system of barter and trade attempts to equate value across goods and services (Appadurai, 1986). This system works when both parties in the exchange agree that the goods or services are of equal value. Agreeing on such a value depends on the subjective value assigned to both goods and services by both people in the exchange. However, depending on subjective value exclusively in this context is problematic. For instance, suppose that one's goods and services are highly valued, rare, and only a few persons are capable of performing the service or supplying the good. The subjective value of the goods and services is high in the persons with demand, but it is relatively lower in the persons with supply. If we factor in deception and mental states, there is a great opportunity for either person in the exchange to be exploited—to pay more for goods and services than these are subjectively worth or to receive less for goods and services than their subjective valuation. However, exploitation as I have just described it is not considered as such by economists: they view such exchanges as generating economic surplus. If, however, such exchanges drained overall value from the economic markets, this would be considered suboptimal because it would lead to inaccurate valuation of goods and services. But there is another, more immediate matter of relevance—defining *subjective*.

Economists advise that there is no such thing as an objective value of goods and services, only the value that people are willing to pay for goods and services: a subjective value. Consequently, the value represented by market prices (or willingness to pay) is really an aggregation of subjective values. Such a definition of market prices relies on consensus. If most people's subjective opinions agree, then that opinion may be considered the actual value (for our purposes). It is very difficult to determine such an objective opinion or value for goods and services in a barter and trade system because goods and services are never exchanged for the same good or the same service. Differing subjective valuations of different goods and services lead to a variety of other goods and services values that could be assigned to the original goods and services. A common value metric is needed: money. Money solves the problem of losing value in the market because it optimizes valuation processes by providing a common metric, allowing the easy comparison of a variety of goods and services. Now value is determined purely by demand and supply on a metric that generalizes across goods and services. Therefore, instead of obscure inferences of a variety of actors' subjective valuations, markets attempt to aggregate those inferences.

Social science researchers have espoused the idea that monetary and social contexts elicit very different behavior. Building on Alan Fiske's (1992) relational theory, they posit that the four types of social relationships—communal sharing, authority ranking, equity matching, and market pricing—can be divided into monetary and social contexts, with monetary contexts comprising market pricing relationships and social contexts comprising the remaining three social relationships (Heyman & Ariely, 2004). Cues within the context specify whether the context is social or monetary, thereby guiding behavior. For instance, in one study participants were offered candy for sale or told that they were free to take the candy. Participants bought multiple pieces of candy when it cost as little as one cent, but they only took one piece in the free condition, presumably because this behavior was in keeping with social norms. Such results are consistent with flexible social cognition, suggesting, at least, that a different kind of cognition is required in the social and monetary contexts.

Additional evidence that differentiates social and monetary contexts comes from a field study of Israeli day-care centers (Gneezy & Rustichini, 2000). In an effort to discourage late-arriving parents at the end of the day, the researchers instituted a monetary fine for late arrivals. Instead of discouraging the practice, the fine increased the number of late arrivals. This occurred presumably because a late arrival before the fine was instituted violated social rules (by inconveniencing the teacher who had to remain behind after hours until the delayed parent arrived), but adding the fine put a price on late arrival, thereby making it a purchasable service. People were willing to pay for this service; therefore, more people sought to exploit this new commodity. Interestingly, the deterrence hypothesis (see the next chapter for a more detailed discussion) predicted that the fine should be effective in decreasing late arrivals if everything else remained unchanged. But the results of the study suggested that the fine did not leave everything else unchanged. Further evidence for this was found when the fine was removed: the number of parents who arrived late remained high. Again, this presumably occurred because the fine had somehow changed the social context: parents were possibly no longer concerned that they were inconveniencing the teachers since the teachers were willing to add a price to their inconvenience. Removing the price did not reintroduce the notion of inconvenience, and the new assumed social norm remained in place.

The 2007 Financial Crisis

An example of how economic contexts interact with flexible social cognition comes from the financial crisis of the first decade of the beginning of the twenty-first century. Some post hoc analyses have identified five points as causal reasons for the collapse: (1) financiers, (2) regulators, (3) the macroeconomic context of moderation, (4) a savings glut in Asian markets that reduced global interest rates, and (5) consumer overborrowing (bad financial decisions). Although some investment bankers who devised Ponzi schemes obviously used deception and possessed criminal intent, many financiers simply behaved in a manner consistent with the economic context. For instance, policy handed down from bank management emphasized the awarding of subprime mortgages (1). These mortgage assets were pooled by the banks to minimize perception of their risk and then were rated by credit agencies, who were employed by the very same banks (2). These same banks then bought the best rated of the assets. These assets were attractive because of low interest rates worldwide since they promised a safe investment with a moderate return (3). Low interest rates, partially caused by vagaries in the Asian economies (4), also promoted purchases of riskier assets that offered higher returns. This period of great moderation suggested low volatility in the markets, justifying bigger long-term risk. When the housing markets eventually tanked, the rest became history (*The Economist*, 2013). Additionally, bad decisions by consumers perhaps were predicated on the idea that their future selves would be able to better handle financial burdens than their current selves (5), a bad prediction that contributed to the collapse.

Some economists have theorized that "animal spirits" present in the markets infected investors causing them to be run by their emotions and to make irrational decisions (Akerlof & Shiller, 2010). This explanation places focus on anomalies in people's behavior, not financial markets, and is consistent with the notion that the social context promoted behavior that required flexible social cognition, which in term facilitated the behavior at each of the five points that led to the crisis. A number of causal factors for the financial crisis highlighted above can be categorized as contextual. The period of great moderation and low global interest rates both suggest a specific economic behavior: an increase in long-term risk. This type of behavior itself is not responsible for the financial crisis, but it occurred in a context where value was misattributed to assets. Without proper ratings of the risk of assets like subprime mortgages, increased long-term risky behavior became detrimental to the market. Similarly, without financial regulation to prevent obvious conflicts of interests that led to rating agencies rating those very assets in which they held a stake, or that allow a policy of subprime mortgage lending, there is little chance that the suboptimal financial behavior can be curtailed. But where specifically did flexible social cognition and a resulting dehumanized perception come into play? As mentioned, flexible social cognition results in a human–nonhuman distinction. If the nonhuman categorization is made, then people are engaging a dehumanized perception that facilitates behavior not reserved for people. In order to approve subprime mortgages, a lender has to not consider the borrower's poor credit history and inability to repay the loan. The lender may not forecast the distress the person may face when her home must now be repossessed and must thereby engage in dehumanized perception. Instead, lenders are focused on the commission received from initiation of the loan and the financial benefit to their company. The person who borrows is considered a "client," not a human being, despite the agency displayed by a human being when taking out a subprime, high-risk mortgage. Similarly, allowing credit rating agencies and banks to be outcome dependent creates a conflict of interest as it ignores the consequences of such relationships for other people, focusing instead on the profit to be made for the lending institution.

Labor Markets

A second example of how economic contexts promote flexible social cognition and the resulting dehumanized perception comes from labor markets. People have been bought and sold throughout most of recorded human history. This practice is rational since people are a valuable resource: their physical and cognitive abilities allow them to complete tasks no other agent can complete. However, this practice presents a paradox: if human beings are indeed a resource that can be subjected to economic market forces, then they can be exploited just as any other resource. But social conventions, norms, and moral rules prohibit exploitation of other people. How then is it possible that people have treated human beings as resources, subjecting them to market forces and often exploitation? Here, flexible social cognition is vital. Having the perceptual flexibility to view some people in some contexts as full human beings, deserving of protection with strong proscriptions against exploitation, and others as not fully human, capable of being bought, sold, and traded, allows some people who serve as resources in economic contexts to be exploited, resolving the paradox.

Although countless historic and modern anecdotes support this line of reasoning, there exists little scientific evidence to support these claims. In one test of this hypothesis (Harris, Lee, Capestany, & Cohen, 2014), participants played a time estimation game, estimating the length of time intervals. Demographic information from these participants was recorded and compiled in a database along with the participants' time-estimation performance, which was normally distributed across the sample. A price was attached to each participant based on her or his overall time-estimation accuracy. A second sample of participants then viewed this information. Endowed with $20, each sample 2 participant (henceforth known as owners) was instructed to use this money to purchase five sample 1 participants (henceforth known as players) to comprise a fantasy team. One to two weeks later, these owners reported to the MRI scanner, where they viewed the pictures of players owned and not owned, as well as their performance. Reward in this MRI component of the study was tied to the "owned" players' performance, such that accurate time estimations led to winnings for the owner, and inaccurate performance did not. After each player's time estimation on one round, the owners revalued the player, deciding if that player's price should increase, decrease, and if so by how much, or simply remain unchanged.

Revaluation behavior suggested that ownership mattered; owners decreased the value of their players after inaccurate performance and increased the value after accurate performance. Here, changes in value may be interpreted as a type of punishment or reward for the players' performance. In contrast, owners decreased the value of unowned players after accurate performance, and increased their value after inaccurate performance. This behavior is more difficult to explain without considering the brain data. When owners first looked at the players' pictures at the beginning of each trial before performance, they decreased activity in their social cognition brain network (engaged dehumanized perception) for players they owned relative to players they did not own. It should be noted that this study is agnostic about whether this decrease was intentional or not, but it occurred in the context of this experiment. Also, it is important to note that reduced social cognition processing does not mean the perceiver does not think the target is human. However, it does mean that in this specific instance, the perceiver did not engage psychological processing typically reserved for human beings. There is a temporal component to consider, and given flexible social cognition, the same target can both engage and fail to engage social cognition in the same interaction or encounter. Further research is required to demonstrate this temporal component. So of course someone will not likely report that a target who is obviously human is not human, but that does not mean the target was not dehumanized in a specific instance or moment to facilitate behavior.

If indeed social cognition gets in the way, hampering decision making during social economic tasks, then this dehumanized perception may be akin to a regulatory strategy, thereby reducing the human exploitation paradox and facilitating the revaluation behavior. Indeed, the decreased activity in social cognition brain regions predicts later revaluation of owned players. In contrast, parts of the brain constituting the reward and subjective value network such as the medial orbitofrontal cortex (MOFC) predicted revaluation behavior for nonpurchased players, suggesting that owners continued to rely on these processes when considering nonpurchased players. This double dissociation supports the hypothesis that social cognition may impede social decision-making in economic contexts.

Additionally, Karl Marx (1968, as cited in Mandel, 1971) raised the idea of laborers being subject to a dehumanized perception when he described proletarians. Specifically, he described types of alienation for such people as a result of being a mechanistic part of a social class structure: alienation from the product of the labor, alienation from the act of producing, and alienation from the self. This latter form of alienation, termed *Gattungswesen* or "species-essence" alienation, can be interpreted as a form of self-objectification. Thus, philosophers have long considered the fact that working within a system, mechanistically producing goods and services, can lead to a perception of the self as less than human.

Labor markets exist today in most industries including the sports industry. This example labor market, like all labor markets, displays a correlation between ability (both physical and intellectual gained through experience) and overall performance. The transparent nature of these markets in sports leagues makes them very attractive for ordinary people to participate, and today, many fantasy leagues exist where ordinary people can trade players who constitute their teams. Real-world performance leads to rewards in these leagues; therefore, many people are invested in the performance of these players regardless of whether they support a particular team or not. Such behavior suggests that people find interacting with labor markets rewarding, perhaps explaining the prevalence of labor markets among other types of economic markets.

However, real-world athletes who have experienced the vitriol of fantasy league owners and fans after a poor performance, or a failure to compete because of injury, have quite readily identified their dehumanized status. For anecdotal evidence, the National Football League's Houston Texans running back Arian Forster complained in 2011 that he was being dehumanized by fantasy league owners after he suffered an injury that prevented him from playing. Many believed that the injury was not severe enough to keep him off the field, and many fantasy league owners who had him on their team took to social media to vehemently criticize Arian Forster for his lack of courage, among other things. It is remarkable to think that an observer would know the extent to which a player suffers from an injury. It is even more remarkable to think that such observers could go to such lengths, thinking about their fantasy teams and the losses they would incur as opposed to the player's actual well-being. Similar claims have been made by athletes in other disciplines, including Derrick Rose of the Chicago Bulls of the National Basketball Association during his lengthy rehabilitation from double knee surgery in 2012 through 2014. Incidentally, we are now in an era when concussions have been identified as the cause of many lingering health problems, including chronic traumatic encephalopathy (CTE), after athletes complete their playing careers. Perhaps dehumanized perception can explain how such a serious injury is not given the attention it deserves and how such athletes continue to be thrown back onto the field of battle after suffering head trauma. Perhaps no one considers their mental state, only the value they bring to their teams.

Final Word on Economic Contexts

Perhaps the most striking conclusion when considering social cognition and economic contexts is that social cognition can get in the way of economic decision-making, biasing it away from the optimal, financially lucrative, decision. This suggests that economic contexts, although created to facilitate division of labor within social groups, may require flexible social cognition to avoid suboptimal decisions. Although economists have long decried the lack of rationality in human behavior, it is difficult to consider social cognition as responsible for such lack of rationality. In fact, social cognition encompasses very logical, rational thinking (see chapter 6). And although older theories of economics might have considered behavior informed by social cognition as irrational, today many economic theorists take account of the social context and social cognition during economic decision-making. Despite the fact that economic exchanges involve people, thinking about the minds of such people can at times be maladaptive in these contexts because people's behavior is driven by economic, not social forces during economic exchanges.

Economic forces provide a different set of rules that govern behavior. Economic markets themselves are a fiction created by economists to better model and explain very social behavior involving economic exchange. Profit-maximizing firms often exploit social cognition to make profits (Akerlof & Shiller, 2015). Self-interest, competition, and exploitation are encouraged and often result in optimal decisions for profit maximization. These motives often contradict social motives, which encourage thinking about other people's interests, cooperation, fairness, and justice. So how is it possible that a social species created such a nonsocial context to facilitate its social behavior? Should there be strict regulations governing behavior in economic contexts given these differences, and should such regulation abide by a social contract? Has the brain evolved separate processing for economic behavior? Such difficult questions remain unanswered.

References

Adams, J. (1965). Inequity in social exchange. In L. Berkowitz (Ed.), *Advances in experimental social psychology* (Vol. 2, pp. 267–299). New York: Academic Press.

Akerlof, G. A., & Shiller, R. J. (2010). *Animal spirits: How human psychology drives the economy, and why it matters for global capitalism*. Princeton, NJ: Princeton University Press.

Akerlof, G. A., & Shiller, R. J. (2015). *Phishing for phools: The economics of manipulation and deception*. Princeton, NJ: Princeton University Press.

Appadurai, A. (1986). Commodities and the politics of value. In A. Appadurai (Ed.), *The social life of things*. Cambridge, UK: Cambridge University Press.

Bobocel, D. R., Kay, A. C., Zanna, M. P., & Olson, J. M. (2010). *The psychology of justice and legitimacy*. New York: Psychology Press.

Fiske, A. P. (1992). The four elementary forms of sociality: Framework for a unified theory of social relations. *Psychological Review*, 99(4), 689.

Gneezy, U., & Rustichini, A. (2000). A fine is a price. *Journal of Legal Studies*, 29(1), 1–17.

Harris, L. T., Lee, V. K., Capestany, B. H., & Cohen, A. O. (2014). Assigning economic value to people results in dehumanization brain response. *Journal of Neuroscience, Psychology, and Economics*, 7, 151–163. doi:10.1037/npe0000020.

Heyman, J., & Ariely, D. (2004). Effort for payment: A tale of two markets. *Psychological Science*, 15(11), 787–793.

Lee, V. K., & Harris, L. T. (2013). How social cognition can inform social decision making. *Frontiers in Neuroscience*, 7, 259.

Mandel, E. (1971). *The formation of the economic thought of Karl Marx: 1843 to Capital*. New York: Monthly Review Press.

Marx, K. (1968). *The German ideology*. Moscow: Progress Publishers.

Raven, B. H. (1999). Influence, power, religion, and the mechanisms of social control. *Journal of Social Issues*, 55(1), 161–186. doi:10.1111/0022-4537.00111.

Rilling, J. K., & Sanfey, A. G. (2011). The neuroscience of social decision-making. *Annual Review of Psychology*, 62, 23–48.

Rosenthal, L., Levy, S. R., & Moyer, A. (2011). Protestant work ethic's relation to intergroup and policy attitudes: A meta-analytic review. *European Journal of Social Psychology*, 41(7), 874–885. doi:10.1002/ejsp.832.

Silberberg, A., Roma, P. G., Ruggiero, A. M., & Suomi, S. J. (2006). On inequity aversion in nonhuman primates. *Journal of Comparative Psychology*, 120(1), 76.

The Economist (2013, September 7). The origins of the financial crisis: Crash course. *The Economist*.

9 Social Contract

> We the People of the United States, in Order to form a more perfect Union, establish Justice, insure domestic Tranquility, provide for the common defense, promote the general Welfare, and secure the Blessings of Liberty to ourselves and our Posterity, do ordain and establish this Constitution for the United States of America.
>
> —Preamble to the Constitution of the United States

Legal systems interestingly blend social cognition with the social context by considering social groups. For instance, one might argue that laws were created to protect in-group members. Rousseau's (1762/2003) theorizing about a social contract works only if people agree to live together. Although governments, militaries, and police forces were created to enforce the terms of the social contract, they do so only for members of their society. Attempts at international armed forces rely on the cooperation of individual countries—national social groups. Therefore, despite Rousseau's theorizing about all humanity, legal systems assume that members governed within a particular jurisdiction belong to the wider in-group. Closer examination suggests that Rousseau's theory may be accurate. As discussed previously, flexible social cognition results in judgments of human–not human that may have evolved in a social context where "human" equaled "in-group member." Since flexible social cognition is involved in the legal context, then it is used to determine who is deserving of being held to the laws of a society and who is not.

The law punishes bad minds that have committed bad behavior. It may sentence people to jail or subject them to fines, but these forms of punishment are motivated by the underlying mental state of the actors in the criminal justice system. As such, the legal context is primarily concerned with mental states and outcomes of behavior. Judgments of culpability, responsibility, intention, and so on feature prominently in legal reasoning (Hoeflich, 1986). Legal decisions (at least in adversarial systems) take into account the mind of the accused, the victims, witnesses or experts brought forward to provide evidence for either side, and other actors in the legal system such as police officers, prosecutors, judges, and members of the jury. This is a feature of the legal system consistent with its concern for the in-group.

The legal system enforces rules that govern social interactions and interactions among individuals, institutions, and entities such as the environment. The key concepts in the preceding statement are "enforces" and "rules that govern social interactions." Note: I consider the latter interactions between entities and the like to be social interactions as well. "To enforce" is a concept that often implies punishment (except for cases where contracts and the like are enforced), and indeed, legal systems are primarily concerned with punishing, not with rewarding appropriate behavior. Therefore, appropriate behavior—behavior consistent with rules that govern social interaction—is expected, and deviation from that norm is punished. Given that the social contract is between in-group members, punishment seems to be a harsh way to deal with breaches. However, one must consider the functions of punishment in order to understand why this is the manner in which groups attempt to govern.

Punishment

Punishment is communicative: it suggests that the victim (the group) is not to be tampered with, and the social contract is not to be breached (Darley & Gromet, 2010). Milder forms of punishment function to communicate these facts to the offender. All forms of punishment, including extremely severe forms of punishment such as life incarceration and death sentences, function to communicate these facts to everyone, including existing and potential in-group members. Therefore, punishment functions as a deterrent within the group and the larger societal context.

Some opponents of the death penalty, the ultimate punishment, argue that this form of punishment does not deter crime. The motive or reason behind a particular form of punishment may be multiply determined, and therefore it may not have the same outcome for every person: some people may be deterred by the death penalty, others may not. Nonetheless, death deters the person sentenced from committing further crimes since the person is no longer alive and able to commit any crimes, suggesting that deterrence is a factor even in this extreme form of punishment.

Punishment is retributive (Carlsmith & Darley, 2008); it allows the group to exact vengeance on offenders (although all forms of retribution do not involve revenge, such as gaining monetary compensation for physical abuse). But how can a group exact vengeance? Since a group is a collection of people, the members could not all strike a blow. Earlier forms of human punishment suggest this was not always the case, as members of a village pelted the offending person in stocks or similar confines. In modern societies financial penalties are often imposed on the guilty, filling public coffers. However, punishment also allows retribution by individual group members because punishment is in the domain of the mind and mental states. Knowing that the guilty have been brought to justice—that they know they cannot breach the social contract without consequences, that the group can police itself for such breaches—arguably satisfies the need for vengeance among individual group members.

Moreover, the desire to punish is an affective response that depends on the amygdala to determine punishment magnitude (Treadway et al., 2014). This affective response, such as anger, also motivates vengeance (Seip, van Dijk, & Rotteveel, 2009) and puts all these mental constructs in place, resulting in retribution, perhaps triggering affective responses consistent with satisfaction. Together, these two functions enable restoration of the victim/group. Therefore, punishment serves to build the cohesion of the group, making it stronger (Homans, 1961; Horne, 2001).

Some researchers argue that punishment has evolved beyond biological systems that promote cooperation and prosocial behavior in humans (see Cushman, 2015, for one such discussion). For instance, institutional forms of punishment are not always motivated by communicative and retributive reasons but, rather, by cultural norms that govern behavior. However, one may ask from whence such norms arose. A parsimonious answer is that these norms evolved from the basic motives described above. A more complex answer views punishment as a way of influencing the future behavior of other individuals. As a result, norms are created, and punishers are professionalized within institutional structures. This more complex answer could be interpreted as an alternative explanation for communication and retribution, although there may be novel predictions from this alternate theory (see Cushman, 2015). Nonetheless, both explanations of punishment motives surround maintaining group cohesion.

Do You Remember the Days of Slavery?

A social contract suggests that all human beings are governed by a set of rules and, as such, are entitled to the same basic inalienable human rights and justice: equal rights and justice. Therefore, if human beings are to have those rights violated, those persons have to be considered not human in order to facilitate those violations. Take for instance the human right of freedom; four hundred years of New World slavery suggests that the captured Africans could not have been considered human.

In fact, the United States constitution states that much: people who were not free (primarily people of African and Native American descent) were considered three-fifths human. Ironically, this value represented a compromise between Southern and Northern states: Southern states wanted each enslaved person to represent one person in order to reflect a higher population of people in their states, whereas the Northern states wanted enslaved people to count as zero, to reduce the population of Southern states, thereby reducing the influence such states would have in Congress. The resulting legal stripping of

40% humanity compounded already entrenched violations of the social contract, which restricted basic freedoms and rights such as literacy, voting, property, and humane, nonviolent treatment. Of course, slavery had been in practice before the formation of the United States, but the American founding fathers found it necessary to specify the reduced humanity of some people in the legal code in order to control power allocated to Southern states. Since the social contract, or law, reflects the broad beliefs of the society in which it governs behaviors, it certainly suggests that this dehumanized perception of enslaved people was indeed prominent at the time in both Northern and Southern states of the United States and perhaps the wider known world.

But the interesting debate does not surround whether these people suffered a dehumanized perception, the result of flexible social cognition (see chapter 8 for empirical evidence that buying and selling people induces dehumanized perception) but, rather, why it was necessary to find flexibility for legal maneuvering among the enslaved people. Perhaps it was deemed important to reflect popular perceptions, or it was important to specify for whom the founding laws were created. Whatever the underlying reason, it may only have been possible because such people were already dehumanized, making a reduction of their legal impact possible. Despite the fact that these enslaved people played an important role in the past, present, and future economic prosperity of the land, the laws were meant for their benefit as much as it was for the horses that transported everyone around, or for the trees from which legally freed descendants of these captured people would swing; that is to say, it was not meant for their benefit.

Racism and the Brain

The argument thus far states that the law reflects popular perceptions within a society. Is there evidence that these perceptions still exist? Definitely, social psychology has created a cottage industry demonstrating racial bias against people in America of predominantly African descent. Since race is a social phenomenon without genetic predeterminacy or foundation (most people of African descent in Americans result from a blending of Native American, European, and African peoples, among others), then this perceptual phenomenon is reflected in the brain just as other perceptual phenomena are, and it is subject to the same effects of cognition. Both explicit and implicit racial bias can be pushed around by the social context and participant goals (see Kubota, Banaji, & Phelps, 2012, for a review of the neuroscience of race). Moreover, scientists have found evidence for early visual differences in the perceiving of race, confirming its perceptual characteristics (for instance, see Ito & Urland, 2005).

Social neuroscience research highlights the importance of the amygdala for implicit prejudice responses. Researchers have demonstrated in American participants that darker-skinned and lighter-skinned faces are processed differently in this brain region (Cunningham et al., 2004) and, moreover, that brain activity in the amygdala correlates with an unconscious measure of racial bias, the Implicit Association Test (IAT), and with trust decisions (Brosch, Bar-David, & Phelps, 2013; Phelps et al., 2000; Stanley et al., 2012). The amygdala also shows increased activation to racial out-groups (Lieberman, Hariri, Jarcho, Eisenberger, & Bookheimer, 2005) and even to minimal group members (groups created in the experimental laboratory that are not meaningful in real life) (Van Bavel, Packer, & Cunningham, 2008). These findings suggest that unconscious and instantaneous stereotyping processes influence amygdala responding.

However, because the amygdala is widely interconnected with the entire brain, it can also influence a variety of cognitive processes and behavior. The amygdala is connected to areas in the prefrontal cortex, such as the ventromedial prefrontal cortex (VMPFC), that regulate emotional responses (Whalen & Phelps, 2009). The VMPFC, through connections with the dorsolateral prefrontal cortex (DLPFC), provides access for conscious, controlled regulation of the amygdala. These two prefrontal regions have been widely implicated in cognitive control—the ability to select thoughts and behaviors in relation to goals and motivations (Kouneiher, Charron, & Koechlin, 2009). These prefrontal regions, in particular the DLPFC, have been recognized to have an important role in the regulation of implicit prejudice attitudes (Kubota, Banaji, & Phelps, 2012). These findings suggest that conscious, controlled processing can exert influence over responding in the amygdala.

Moreover, the amygdala supports associative learning of emotionally salient stimuli (Sarter & Markowitsch, 1984). The striatum, a brain region associated with learning and rewards, also has connections to the amygdala (Schultz, Dayan, & Montague, 1997). Signals in the amygdala direct activity in the striatum, tuning this learning mechanism to salient stimuli in the environment (Li, Schiller, Schoenbaum, Phelps, & Daw, 2011). A recent study also demonstrates that emotion increases learning speed during reward learning and further ties this signal to the striatum and amygdala (Watanabe, Sakagami, & Haruno, 2013). Thus, the amygdala also directs learning.

Another critical brain region connected to the striatum and associated with punishment reinforcement learning is the insula (Seymour, Singer, & Dolan, 2007). The insula provides a conscious experience of internal bodily states and responds in tasks associated with disgust and empathy (Craig, 2002, 2009). It also is coactive with the anterior cingulate cortex (ACC), another region implicated in cognitive and executive control during both physical (Sawamoto et al., 2000) and social pain (Eisenberger, Lieberman, & Williams, 2003), providing access to conscious bodily responses. The insula has been implicated in stereotype activation (Harris & Fiske, 2006), and its role in prejudice in concert with the amygdala suggests a complex emotional response. During enactment of prejudiced behavior, social targets trigger fear and disgust responses (among other emotional responses), shaping a particular behavioral repertoire.

"Black Ape"

In addition to implicit bias, researchers have also documented that people of predominantly African descent are still dehumanized in America. Specifically, people of predominantly African descent are associated with an ape stereotype. Before the 1933 movie *King Kong*, such people have often been considered not just subhuman but specifically ape-like. "Scientific" work by American founding fathers such as Thomas Jefferson documented physical differences between Africans and people of European descent, relating these differences as evidence that Africans and people of African descent were more closely related to apes than to humans along the evolutionary continuum. Remarkably, this idea still exists today (the human continuum or "ascent scale" is actually a dependent variable used in dehumanization research, and people today do endorse that other people fall lower on the continuum) and in North America, Brazil, South Africa, and other parts of the world, has been associated with negative legal consequences for people of predominantly African descent.

In one experiment American participants of predominantly European descent were first subliminally primed with either darker-skinned or lighter-skinned faces and then asked to press a button when noisy images of animals were visible. Visibility increased in the images as white noise slowly lifted from them. Participants were faster to identify ape images (images with more noise, or fuzzier images) than other types of images after being primed with darker-skinned faces. This suggests that participants were more visually attuned to darker-skinned faces after the ape prime. In a second experiment participants completed a task where either a darker-skinned or lighter-skinned face served as a cue to the location of a probe (a dot) on the screen (dot-probe task). First, they were subliminally primed with either ape pictures or nonape animal pictures before completing the dot-probe task. Participants were faster to identify the dot-probe when they were primed by the ape images and when a darker-skinned face cued its location. This suggests that participants paid more attention to darker-skinned faces after the ape prime, facilitating their identification of the dot-probe (Goff, Eberhardt, Williams, & Jackson, 2008).

Finally, and most disturbingly, the likelihood of being sentenced to death in Philadelphia was predicted by the amount of "black ape" associations being made in the media to the defendant (Goff et al., 2008). In this experiment researchers analyzed the *Philadelphia Inquirer*'s (the local newspaper) coverage of 153 death-eligible cases (cases where the defendant could be sentenced to death if convicted), noting instances where ape-related words (such as ape, monkey, gorilla) were used to describe defendants of predominantly African or predominantly European descent. Such words were used nearly four times as often for cases involving defendants of predominantly African descent compared to defendants of predominantly European descent and nearly twice as often for defendants of African descent who were sentenced to death than those whose lives were spared. Additionally, the use of such words predicted death sentences, suggesting that dehumanized perceptions are not euphemistic but, although the evidence is only correlational, carry severe real-world consequences.

No Humans Involved

Police are supposed to serve and protect their communities. However, the flexibility in interpretation of the social contract and flexible social cognition mean that the police fail to fulfill those duties for some citizens. People of predominantly African descent in America suffer from dehumanized perception due to flexible social cognition, and incidents of fatal police shootings of citizens of predominantly African descent are not uncommon in America. These continued acts of violence seem akin to days of lynching of Americans of predominantly African descent in the Jim Crow era, although such acts were often not carried out by those in authority (indeed, as recently as 2012 the shooting of Trayvon Martin was perpetrated by a citizen of predominantly European descent who claimed to be protecting his community).

The Jim Crow era may best be remembered for the failed separate but equal propaganda used to justify continued oppression of Americans of predominantly African descent after the Civil War and Reconstruction period. As was the case with apartheid in South Africa, people of predominantly African descent were not allowed to use the same public facilities (toilets, drinking-water fountains, etc.) as people of predominantly European descent, attend the same schools as people of predominantly European descent, eat at the same restaurants as people of predominantly European descent, or participate in civil society such as vote in local and national elections. A common occurrence during the half-century following the Civil War was "strange fruit" hanging from the trees: the bodies of people of predominantly African descent lynched by a mob of people of predominantly European descent for perceived violations of the law (or just the customs) of the era.

The major difference between police violence today and Jim Crow era lynching is that, unlike during the years of Jim Crow, when the law was being taken into the hands of citizens of predominantly European descent, police officers are by law licensed to use force (and to kill as necessary when they perceive a threat to themselves and civilians). However, the prevalence of implicit bias in society suggests that such acts by police officers may not be intentional, although this is subject to substantial debate. In the end, in all cases, the perpetrators of fatal violence against people of African descent are often not subject to the punishment of the law.

These instances of police brutality and abuse of power should elicit widespread policy attempts to repair the social contract to ensure inclusion of Americans of predominantly African descent into its protection. However, dehumanized perceptions may have prevented that from happening. Anecdotal evidence for this comes from police officers themselves; they use terms like "no humans involved" and "gorillas in the mist" in police radio chatter to describe scenarios involving people of predominantly African descent (Goff et al., 2008). Such obvious dehumanization cries out for some type of remedy. Researchers who have taken this challenge study police officers in highly ecologically valid paradigms to determine the psychological mechanisms underlying this phenomenon. For instance, one such study uses virtual reality in the police training environment to measure police officer reactions to encounters with residents who are being mildly uncooperative (Goff & Martin, 2013).

Final Thoughts on the Social Contract

I have described race in the United States because I am familiar with that particular context, given the thirteen years I have lived in the United States and my racial classification as "black" there (despite my rich genetic heritage which includes European, African, and even Jewish ancestry). I was viewed as "black," and yet opening my mouth and speaking with a Caribbean accent spared me occasionally from the label African American. However, the problems I have described above as present in American society are present in all human societies, not just in the United States. In my own Caribbean island nation, social class is used as the dividing characteristic, and people are dehumanized accordingly. In other countries religious affiliation, tribe, and country of origin are all used to carve up the social space and denigrate some minority in favor of a majority (the United Kingdom, where I am currently writing, has experienced a wave of anti-British sentiment following the vote to leave the European Union, leading to increased attacks on people perceived as non-British).

Identifying someone as human bestows on that person the right to be governed by the social contract. Consequently, flexible social cognition plays a key role in determining who is subjected to the protection of society and who is not. In this chapter I have discussed racial bias in the United States as but one example of how social cognition may be withheld and the consequences of such bias for legal decisions. However, the broader point ignores race and addresses whether other people are seen as human or not. I often argue that the best solution to racial bias problems (or any group bias really) is destruction of the categories. For instance, "black" has been associated with negative events, white with positive events since the Manichean philosophy of ancient Greece[1] (Harrell, 1999). Using such preexisting concepts when describing people automatically imbues one group with positives and the other with negatives. Because these concepts, in addition to other, traditional racial stereotypes, are constantly being reinforced, often without conscious awareness, it is almost impossible to eliminate "black" stereotypes.

Removing the social category terms and replacing them, for instance, simply with "American" now categorizes all people as citizens of a place and subject to the same social contract. This would be an improvement on the term "African American" because such a term still dissociates such people from other Americans. I remember sitting in an auditorium at Howard University (a historically black university) in the year 2000, a few weeks into my undergraduate career, at a rally for presidential hopeful Al Gore. His aides had handed out posters with "African-Americans for Gore" emblazoned across the front. Students, on receiving the posters, began to grumble. Suddenly, all the students started folding the posters, making a crease between "African" and "Americans" so that "Americans for Gore" was now visible. A loud cheer erupted from the auditorium as the practice spread among the audience. Such an early experience with race in America drove home to me the point that (at least some) Americans of predominantly African descent want to be considered simply as Americans, and labels like "African American" simply reinforce the historic divide.

Other researchers have argued for similar superordinate identity categorization as a prejudice reduction technique (see Gaertner & Dovidio, 2014, for a collection of essays describing similar points of view). However, the reader may be troubled at this suggestion if she or he thinks that such a move would lead to a disregarding of the culture of people of predominantly African descent in the United States, rendering it invisible. This is especially true since most people in the United States equate "Americanness" with "whiteness" (Devos & Banaji, 2005); therefore, removing the social group labels may result in a "whitening" of the cultural landscape. I argue that this occurs *because* of the current context in which racial group categories are the most salient defining feature of other people. If this were not the case, a superordinate categorization might be effective. But given that this is indeed the case, are a "whitening" of America and a fading of "black" culture a preferred outcome? To me, this question fails to appreciate that "black" culture *is* American culture; simply categorizing it as a cultural subtype misunderstands that a superordinate identity renders such subcategorization useless.

Social categories serve to carve up the social contract, reserving its protections for only some. Because applying the social contract to a person requires acknowledgment of the person as a full human being, withholding social cognition can deny a person the social contract. This most heinous of prejudices violates fundamental human rights, and scientists and policy makers are constantly trying to eradicate such violations. Flexible social cognition seems like a good place to start.

Note

1. The reader may wonder why I do not use the terms "black" and "white" to describe people when referring to race. Such terms further dehumanize both sets of people, reducing them to metaphors, not people (see Rice, 2008, for more examples of caricatures of people of African descent). These metaphors are rooted in ancient Manichean philosophy, where black captures bad, and white represents good. In fact, "black" people are brown or darker-skinned, and "white" people are pink or lighter-skinned. Perhaps people adopted the terms "black" and "white" as a metaphor to reinforce power and perceived moral differences between the two sets of people (see Harrell, 1999). Its continued use today only adds to problems of racial division, and ignores the diverse genetic composition of many "black" and "white" people living in the New World.

References

Brosch, T., Bar-David, E., & Phelps, E. A. (2013). Implicit race bias decreases the similarity of neural representations of black and white faces. *Psychological Science*, 24(2), 160–166.

Carlsmith, K. M., & Darley, J. M. (2008). Psychological aspects of retributive justice. In M. P. Zanna & M. P. Zanna (Eds.), Vol. 40, pp. 193–236). *Advances in experimental social psychology*. San Diego: Elsevier Academic Press. doi:10.1016/S0065-2601(07)00004-4.

Craig, A. D. (2002). How do you feel? Interoception: The sense of the physiological condition of the body. *Nature Reviews Neuroscience*, 3(8), 655–666.

Craig, A. D. (2009). How do you feel—now? The anterior insula and human awareness. *Nature Reviews Neuroscience*, 10(1): 59–70.

Cunningham, W. A., Johnson, M. K., Raye, C. L., Gatenby, J. C., Gore, J. C., & Banaji, M. R. (2004). Separable neural components in the processing of black and white faces. *Psychological Science*, 15(12), 806–813.

Cushman, F. (2015). Punishment in humans: From intuitions to institutions. *Philosophy Compass*, 10(2), 117–133.

Darley, J. M., & Gromet, D. M. (2010). The psychology of punishment: Intuition and reason, retribution and restoration. In D. R. Bobocel, A. C. Kay, M. P. Zanna, & J. M. Olson (Eds.), *The Psychology of Justice and Legitimacy* (pp. 229–249). New York: Psychology Press.

Devos, T., & Banaji, M. R. (2005). American= white? *Journal of Personality and Social Psychology*, 88(3), 447.

Eisenberger, N. I., Lieberman, M. D., & Williams, K. D. (2003). Does rejection hurt? An fMRI study of social exclusion. *Science*, 302(5643), 290–292.

Gaertner, S. L., & Dovidio, J. F. (2014). *Reducing intergroup bias: The common ingroup identity model*. Hove, UK: Psychology Press.

Goff, P. A., Eberhardt, J. L., Williams, M. J., & Jackson, M. C. (2008). Not yet human: Implicit knowledge, historical dehumanization, and contemporary consequences. *Journal of Personality and Social Psychology*, 94(2), 292.

Goff, P., & Martin, K. (2013). Unity breeds fairness: The Consortium for Police Leadership in Equity, Report on the Las Vegas Metropolitan Police Department.

Harrell, C. J. P. (1999). *Manichean psychology: Racism and the minds of people of African descent*. Washington, DC: Howard University Press.

Harris, L. T., & Fiske, S. T. (2006). Dehumanizing the lowest of the low neuroimaging responses to extreme out-groups. *Psychological Science*, 17(10), 847–853.

Hoeflich, M. H. (1986). Law and geometry: Legal science from Leibniz to Langdell. *American Journal of Legal History*, 30, 95–121. doi:10.2307/845705.

Homans, G. C. (1961). *Social behavior: Its elementary forms.* New York: Harcourt Brace Jovanovich.

Horne, C. (2001). The enforcement of norms: Group cohesion and meta-norms. *Social Psychology Quarterly,* 64(3), 253–266. doi:10.2307/3090115.

Ito, T. A., & Urland, G. R. (2005). The influence of processing objectives on the perception of faces: An ERP study of race and gender perception. *Cognitive, Affective & Behavioral Neuroscience,* 5(1), 21–36. doi:10.3758/CABN.5.1.21.

Kouneiher, F., Charron, S., & Koechlin, E. (2009). Motivation and cognitive control in the human prefrontal cortex. *Nature Neuroscience,* 12(7), 939–945.

Kubota, J. T., Banaji, M. R., & Phelps, E. A. (2012). The neuroscience of race. *Nature Neuroscience,* 15(7), 940–948. doi:10.1038/nn.3136.

Li, J., Schiller, D., Schoenbaum, G., Phelps, E. A., & Daw, N. D. (2011). Differential roles of human striatum and amygdala in associative learning. *Nature Neuroscience,* 14(10), 1250–1252.

Lieberman, M. D., Hariri, A., Jarcho, J. M., Eisenberger, N. I., & Bookheimer, S. Y. (2005). An fMRI investigation of race-related amygdala activity in African-American and Caucasian-American individuals. *Nature Neuroscience,* 8(6), 720–722.

Phelps, E. A., O'Connor, K. J., Cunningham, W. A., Funayama, E. S., Gatenby, J. C., Gore, J. C., et al. (2000). Performance on indirect measures of race evaluation predicts amygdala activation. *Journal of Cognitive Neuroscience,* 12(5), 729–738.

Rice, D. W. (2008). *Balance: Advancing identity theory by engaging the black male adolescent.* Lanham, MD: Lexington Books.

Rousseau, J.-J. (1762/2003). *On the social contract.* New York: Dover Press.

Sarter, M., & Markowitsch, H. J. (1984). Collateral innervation of the medial and lateral prefrontal cortex by amygdaloid, thalamic, and brain-stem neurons. *Journal of Comparative Neurology,* 224(3), 445–460.

Sawamoto, N., Honda, M., Okada, T., Hanakawa, T., Kanda, M., Fukuyama, H., et al. (2000). Expectation of pain enhances responses to nonpainful somatosensory stimulation in the anterior cingulate cortex and parietal operculum/posterior insula: An event-related functional magnetic resonance imaging study. *Journal of Neuroscience*, 20(19), 7438–7445.

Schultz, W., Dayan, P., & Montague, P. R. (1997). A neural substrate of prediction and reward. *Science*, 275(5306), 1593–1599.

Seip, E. C., van Dijk, W. W., & Rotteveel, M. (2009). On hotheads and dirty harries. *Annals of the New York Academy of Sciences*, 1167(1), 190–196.

Seymour, B., Singer, T., & Dolan, R. (2007). The neurobiology of punishment. *Nature Reviews Neuroscience*, 8(4), 300–311.

Stanley, D. A., Sokol-Hessner, P., Fareri, D. S., Perino, M. T., Delgado, M. R., Banaji, M. R., et al. (2012). Race and reputation: Perceived racial group trustworthiness influences the neural correlates of trust decisions. *Philosophical Transactions of the Royal Society of London. Series B, Biological Sciences*, 367(1589), 744–753.

Treadway, M. T., Buckholtz, J. W., Martin, J. W., Jan, K., Asplund, C. L., Ginther, M. R., et al. (2014). Corticolimbic gating of emotion-driven punishment. *Nature Neuroscience*, 17(9), 1270–1275.

Van Bavel, J. J., Packer, D. J., & Cunningham, W. A. (2008). The neural substrates of in-group bias: A functional magnetic resonance imaging investigation. *Psychological Science*, 19(11), 1131–1139.

Watanabe, N., Sakagami, M., & Haruno, M. (2013). Reward prediction error signal enhanced by striatum–amygdala interaction explains the acceleration of probabilistic reward learning by emotion. *Journal of Neuroscience*, 33(10), 4487–4493.

Whalen, P. J., & Phelps, E. A. (Eds.). (2009). *The human amygdala*. New York: Guilford Press.

10 Further Societal Implications

Human beings are capable of great evil. Genocide, torture, murder, rape, slavery, and human trafficking all provide examples of the many ways human beings dehumanize other people. But are such behaviors perpetrated by bad people, suggesting human beings are inherently evil, making all this inevitable, or is there another explanation? Perhaps the social context—the social environment, including other people, institutions, and culture—motivates dehumanization. And the social context can be changed. (Harris, 2014)

I have attempted to tell an interesting story about a feature of the mind: flexible social cognition makes visible people's minds invisible (and brings nonhumans to life). Assuming sound logic, the hopefully parsimonious explanations provided have implications for other psychological processes. Stated differently, if small social groups in our evolutionary history promoted flexible social cognition, then how can we accurately characterize the way social context affects both brain activation patterns and human beings' lives together today? The evidence discussed suggests that the effect of the social context on human behavior is complex and that human beings may not be able to live together without cases of massive desolation. In essence, group conflict, prejudice, and inhumane group-based behavior may continue until human beings further evolve and flexible social cognition no longer provides a fitness advantage.

Take for instance the equation I first proposed at the beginning of this book,

$$M_t = f(a, s) + e.$$

One task involves further specifying *a, s,* and *e*. The former refers to a variable that accounts for the person. A particular person, in any kind of circumstance, is going to respond in a particular way. This individual difference is based on biological underpinnings that promoted survival in a particular context. The influence of this particular context is captured by the *s* component of the equation. Therefore, neuroscience approaches that probe brain activity may help to specify $f(a, s)$; this function determines mental tension and can be reflected in corresponding brain activity. The activity can be modeled across a variety of specified *a*'s and *s*'s. Given the extensive work in personality psychology specifying different types of personality variables, specifying the social context remains the challenge. Moreover, explaining the unexplained variance captured by the error term is an even more daunting challenge.

The social context presents a problem because it is so varied. One simple attempt to characterize it describes it as in-group–out-group, but this is a broad-brush approach that collapses across the variability within the two. Furthermore, it undermines the power of the social context to make in-group–out-group distinctions flexible over time. Therefore, this may not be an appropriate way to categorize the manner in which the social context can influence social cognition, brain activity, and behavior. So how then do we properly characterize something as variable as the social context? Psychologists have longed for a periodic table of behavior for quite some time, and specifying the social context may provide such a solution.

The Social Context Specified?

If a social psychologist were to specify the social context, she or he might begin by attempting to collect the variety of social contexts that exist. Such variety is easily found in the social psychology literature and includes studies that explore contextual differences in behavior driven by power, status, group (including race, gender, age, and minimal groups), emotion (including threat and approach motives), similarity, familiarity (experience or learning), social motives, and social role. One possible problem with this approach surrounds the interaction of many of these broad contextual categories with each other and with their own subcategories (having an elderly female subordinate versus young male superior). Therefore, this results in an infinite number of social contexts because each interaction creates a novel social context. Second, the variables described above all ignore the physical context and its impact on behavior; behavior also differs in different places with complete strangers (a party versus a library), and chemical concentrations present in all human societies also exert some influence on behavior—an issue I have ignored in this book. Moreover, these physical contexts interact with the more "social" or human contexts described above, resulting in even more unique social contexts.

Perhaps a more productive approach might be instead to focus indirectly on the social context by defining the function of social cognition: prediction or explanation. People attempt to predict other people's behavior, and these predictions govern their own behavior. This places impression management concerns at the forefront of defining the social context, making base rate or social norms an important component of the social context. Second, attempts at explanation surround the imaginary: behavior results from things that exist only in people's minds. Although predictions and explanations result from logical inferences, they ignore accuracy and serve to satisfy meaning-making concerns that also govern people's behavior respectively. Therefore, this attempt to specify the social context considers base rate and imagination as the pillars of the social context from which social behavior may be explained and *s* specified. Base rates provide the engine, whereas imagination fills in all the rest.

The Evolving Mind

A second implication of the ideas discussed raises the question: How can social behavior evolution continue to occur? Stated differently, how is it possible that human beings move beyond simple human–nonhuman appraisals of others? Perhaps this ability is required for social functioning, but it may not always be necessary. Imagine what would happen if humans truly became the in-group. At the moment, humankind is the dominant species on the planet. Since no others pose a significant threat to our survival, there is no motivation to do away with flexible social cognition because the greatest threat to survival usually comes from another person. If another species, such as dogs or mutated humans, evolved further to challenge humans' preeminence, or if a species were to arrive from beyond the planet to pose such a challenge, only then might flexible social cognition fulfill its purpose and be used to determine human friend from nonhuman foe. This is admittedly the stuff of science fiction, but if this is indeed what will be required, humanity faces a pretty dim future.

But perhaps the dramatic is not necessary. As mentioned above, flexible social cognition is the result of an imagination process. The resulting inferences are not necessarily grounded in reality but are consistent with existing perceptions, social hierarchies, and internal motives. Importantly, imagination is not real and can be altered. What is required may not be the existential threat from another species to drive humans to use human–nonhuman appraisals more appropriately but a changing of the imagination. Interestingly, the threat of another species works because it shifts the imagination; now other humans have become allies, not enemies. However, other things may shift the imagination and resulting perceptions and inferences as well. Evidence for this comes from changing national and ethnic stereotypes over time in United States society. About every twenty years since 1933 psychologists have measured perceptions or stereotypes of groups in nationally representative surveys in American society (Gilbert, 1951; Karlins, Coffman, & Walters, 1969; Katz & Braly, 1933). These studies have now documented shifting stereotypes over almost a hundred-year period. Groups like the Japanese who were once maligned and viewed as cold are now viewed as warm, whereas others, the Muslims and Arabs, who were once viewed as nonthreatening, are now viewed with great suspicion. This provides evidence that perceptions naturally shift over time (though for a counterargument, see Devine & Elliott, 1995), speaking to the changeability of human imagination.

So is the solution simply to wait until the perceptions that shape flexible social cognition shift? Perhaps not, since the groups may change their relative positions, but the psychological phenomenon remains. What may be required instead is a shifting of the phenomenon. Flexible social cognition presumably evolved to support social function at a time when people lived in relatively small, insular social groups. Modern humans live in a vastly different social world, one where people are instantly connected anywhere on the planet, and information about a person is readily available from a variety of sources before one even encounters them physically. This very recent change in human social evolution should render flexible social cognition useless. Indeed, psychologists and philosophers have argued that human beings are becoming less antisocial as a species, perhaps due to the rise of logic and reason (Pinker, 2011). Such increased use of logic and reason pushes moral definitions, extending boundaries of moral protection beyond humans to other animals (in the case of vegetarians) (see Singer, 1975) and maybe even the planet.

To clarify, morality was once believed to be composed primarily of emotional responses to harms. This more deontological perspective was challenged by a more utilitarian perspective that describes morality as the result of cognitive calculations that weigh gains and losses before arriving at a moral decision. This latter conceptualization of morality places a burden on cognitive faculties, specifically logic and reason. Therefore, some researchers such as Steven Pinker now conceive of morality as the result of intelligence.

However, this argument, if valid, should extend to prosocial behavior as well. Has modern humanity decreased its humanitarian efforts? Steven Pinker would argue that these efforts have increased. Therefore, if flexible social cognition underlies both pro- and antisocial behavior, and modern social life is rendering it useless, then we should see a decrease in both pro- and antisocial behavior (unless of course the fall of flexible social cognition increases the use of another psychological process that promotes prosocial behavior). But this is not the case, suggesting that flexible social cognition may continue to find purpose in modern human society. However, its usefulness may have expanded, rather than contracted; since flexible social cognition is no longer necessary to categorize strangers into friend or foe camps, it may be used instead to rob a person of his or her humanity when it is advantageous for an agent to do so. We have discussed cases in economics where flexible social cognition led to appraisal of people as less than human, thereby triggering increased economic performance. The same may also hold true in medical domains and competitive group contexts. In essence, flexible social cognition is still a useful pro- and antisocial tool.

Medical Care

Medical professionals face a difficult task: on the one hand they must be extremely compassionate and empathic with the goal of proper diagnosis and treatment as well as comforting to patients and family members. On the other hand, they are often required to do things that we do not typically do to other people, such as slice them open. The latter task may require blurring the human–nonhuman distinction, viewing the person's body as a machine that needs fixing. Therefore, both tasks require different levels of empathy. How might medical professionals be able to accomplish both? Here, flexible social cognition may be key. Because it results in a quick, flexible appraisal of the agent as human or not, it is able to hold a humanized and dehumanized perception of the same person, but at different times in different contexts. Let us consider the case of the surgeon (it should be noted that most doctors are not surgeons, but the general process may apply to them and to other health-care providers as well) who may humanize the patient in her office when consulting on a treatment plan, course of surgery, and recovery schedule, drafting plans that best fit the individual medical, financial, and social needs of the patient. The same surgeon may then dehumanize the same patient when in the operating theater, viewing the triple bypass procedure as necessary to fix a broken object, suppressing concerns surrounding the patient's mental state. The example above is not bad, that is, the resulting excellence in (1) surgery achieved through precision that requires regulating social cognition and (2) the pastoral care achieved through empathy and compassion for a patient are the gold standards of health care. Therefore, flexible social cognition can lead to very positive outcomes, not simply to genocide, racism, and harm.

Initial evidence suggests this may be the case; research in the medical decision-making literature describes worker burnout due to a failure of compassion (Maslach, 1982), and medical professionals with higher dispositional empathy levels are more likely to suffer burnout (Vaes & Muratore, 2012). These findings suggest that high levels of empathy may not always lead to good things, and flexible social cognition may aid in prosocial behavior.

Competition

A remarkable feature of human culture is the large, competitive event. Whether this involves gambling, sporting events, or pageants, human beings enjoy competing. A parsimonious account suggests winning is fun. Indeed, brain-imaging evidence suggests that people enjoy winning; the striatum, which responds to food and monetary rewards across species, responds to winning in competitive contexts such as auctions (Delgado, Schotter, Ozbay, & Phelps, 2008). This suggests that competitive contexts continue to exist because they are intrinsically rewarding; simply ask any fan at the World Cup or Olympics. I should note that I am not saying that competitors reduce the opposition to flesh and bones; in fact, there is substantial social cognition involved to infer intentions and other mental states necessary to win in a competitive context. However, competitors do not care for the opposition to have positive outcomes and may reduce social cognition when such outcomes can be considered.

At this point it is important to focus the discussion on competition by dissociating different competitive contexts and restricting our discussion simply to the appropriate contexts. There are intergroup and interpersonal competitive contexts. This is akin to individual and team sports, although most athletes who compete in individual events also represent teams at international competition, illustrating the relevance of groups in competitive contexts. There are also competitive contexts against other people and competitive contexts against probability. One can argue that Bayesian processes determine the outcome in both contexts, but they may be perceived differently.

It is useful to compare competition in human and nonhuman species. Competition enables pursuit of two related yet independent social motives: status and power. Power is defined as resource control (Guinote, 2007), whereas status evaluates one's relative social standing or value within a social group hierarchy (Fiske, 2010). Of course, individuals with control over resources are highly valued within their social groups, and highly regarded others often control resources through their influence, but both social factors can be decoupled. Interestingly, fierce violent intergroup competition is predominantly over resources, whereas more friendly competition is often about relative standing. In nonhuman animals, status and power are conflated, begging the question why they are dissociable in humans. One reason may surround the sheer size of human groups. Considering simply ethnic groups involves hundreds of thousands of group members. This allows for smaller groups within larger groups, which creates more opportunities for high relative standing but not more resources to control. As a result, relative social standing has become more flexible, decoupled from resource control in humans.

Moreover, mental state inferences play a central role in both resource and relative standing arenas. In regard to power differences, social psychological research suggests that people in positions of low power individuate (think about the mental states of) high-power others, whereas people in high-power positions fail to individuate others, instead relying more on stereotypes (see Fiske, 2010, for review). Similarly, phenomena such as inequity aversion are found across species (as discussed in chapter 2), suggesting that even other species are sensitive to imbalances in distribution of resources. Now the question arises whether social cognition is necessary to display behavior consistent with experiencing inequity aversion. I argue that flexible social cognition is necessary.

Consider the case in which one monkey receives cucumbers for performing an instrumental action while another monkey receives grapes. The first monkey prefers grapes to cucumbers. If the first monkey displays negative affect on detecting the difference in reward for the same behavior and refuses to engage in the behavior further, essentially going on strike, then one may infer inequity aversion. But does this first monkey think that the second monkey prefers grapes as well, or is it sufficient that grapes are available as an alternative reward and simply were not given? Researchers have investigated this and similar questions (see Brosnan & De Waal, 2003; Chen, Lakshminarayanan, & Santos, 2006). Monkeys are also sensitive to whether the high-value reward was given to another monkey or placed in an empty chamber (Chang, Winecoff, & Platt, 2011). This suggests that the inequity aversion is in response to another agent getting a high-value reward for the same effort. Without knowing that the other monkey considered the reward high in value, there would be no monkey strike. Status also requires mental state inferences. Since status is by nature comparative, as is the inequity aversion described above, it has to take into account the mental states of the other. The emotions that result from status comparisons such as envy and *schadenfreude* support this assertion.

The Next 100 Years

Human beings have developed technology at a remarkable rate. The most recent advances have furthered our ability to communicate with other people and avoid menial tasks. However, the latter type of technological advancement is still ongoing, and a day may come within the next century when humans will no longer require low-income workers because machines will have become capable of performing menial tasks. Will such a day eradicate low socioeconomic social groups? Perhaps not, at least, not if flexible social cognition is still in widespread use in human society. So is there any technological advance that can eradicate flexible social cognition? The answer may lie in the first type of technological advancement: communication.

The invention of the Internet and subsequent social media has made the context of other people's minds extremely accessible. At no point in human history has it been as easy to find out what someone else is thinking. Moreover, we now have access to such information from a vast number of people. Such advances may make flexible social cognition all the more necessary because of the sheer number of people who can now be tracked. My intuition suggests that it may have the opposite effect: the abundance of mental state information may make the human–nonhuman appraisal useless. However, it is possible that such technologies may further polarize social groups if people just explore the mental states of their own group members.

Coupled with communication and technological advance is the rise of artificial intelligence (AI). Though people today may still treat AI as nonhumans, it seems likely that in the next century, the information coming from AI and other humans may be seamlessly blended because people will no longer have the same amount of face-to-face interactions with other people on the Internet. Therefore, despite the preponderance of mental state information, it will not be accompanied by the physical social interaction in the way such interaction has always occurred in human history. Together with the rise of AI, human beings may find the human–nonhuman appraisal less useful in the service of determining with whom one would like to interact. Or perhaps online interactions present a new set of threats that the brain has not evolved to detect or respond to.

The science fiction of the previous paragraph aside, one possible way of curbing the harmful effects of flexible social cognition may be by focusing on the shared biology of humans. Specifically, human beings have more genetic similarity to each other than to other species. All human beings share a common ancestor in the more recent past than they share with other species. As a result, human beings can identify as a group with each other more so than they can identify as a group with other species. This suggests that a perception of a common humanity is plausible. Today, people predominantly discriminate social groups using physical features (although not always; for instance, religious and tribal affiliations are used quite often), suggesting underlying genetic differences. In fact, people have historically used physical feature differences to justify why some people are less human than others, such as the work of Thomas Jefferson and others to scientifically "prove" that Africans were less human that Europeans as discussed in chapter 9. These attempts at identifying differences are used to construct narratives that explain (and justify) who should be treated as human beings.

Narratives are powerful tools that help create senses of self and community, for reasons related to explanation: they fulfill the need for sense-making discussed in chapter 4. This psychological phenomenon influences perceptions, attitudes, and emotional responses and helps maintain group differences. However, these narratives are often inaccurate and based simply on people's biased points of view (a feature of all humanity). Correcting these narratives, replacing them with more accurate descriptions, is a goal of reconciliation commissions. However, the truth always remains subjective, and the degree to which people believe one person's point of view is always going to be influenced by who that person is and how the truth fits with the narrative people have grown accustomed to for reasons related to consistency.

One possible solution to this problem is biology. Despite the nefarious use of biological features of human beings to promote group differences and to exclude some humans from the category of humanity, biology does hold the promise of underlining the common similarity of all humans. For instance, human beings have had a checkered groups history, where certain groups have been victimized, their group members scattered, depleted, and collective histories destroyed. Additionally, human beings have moved around the planet extensively, migrating between existing social groups and forming new ones. The only complete record of this extensive social group upheaval is us: people today who are the descendants of these migrating people. Although historical records (immigration roles, ship manifests, etc.) provide some information to track this migration, the genetic information contained within each of us provides an accurate record of this upheaval. Mapping this information might show the commonality present in all of us. Given that social groups are constructed, such evidence might challenge these constructions, relegating flexible social cognition in intergroup contexts to the past.

An alternative, more pessimistic, view comes from the work on biological essentialism, much of which highlights the negative consequences of perceiving groups in biological terms (see, e.g., Brescoll & LaFrance, 2004; Dar-Nimrod & Heine, 2011; Jayaratne et al., 2006; Williams & Eberhardt, 2008). This work suggests that it might be possible that perceiving one's in-group in biological terms might lead to even more dehumanization. For example, if people of predominantly European descent already perceive people of predominantly African descent in animalistic terms, is it possible that focusing on biology would lead them to conclude that people of predominantly African descent are even less human-like (i.e., not "I share so much biological material with people of African descent, so they must be fully human just like me" but "if all humans share much biological material in common, people of African descent must have different types of genes than I do, since they are not nearly as human as I am"). People may not be ready to give up dehumanization altogether, as opposed to finding ways to turn positive interventions into ways to dehumanize more. I hold greater hope for humanity and cannot believe such claims.

Final Thought

In conclusion, this discussion raises the question whether humans are actually better off without flexible social cognition. Although I cannot answer this because of the highly subjective nature of the phrase "better off," I can say that if flexible social cognition evolved to aid in survival and reproduction, then perhaps we can focus on social contexts where such flexibility renders positive outcomes. If humanity can leverage the ability for these purposes, perhaps humanity can be better off because of it.

References

Brescoll, V., & LaFrance, M. (2004). The correlates and consequences of newspaper reports of research on sex differences. *Psychological Science*, 15(8), 515–520.

Brosnan, S. F., & De Waal, F. B. (2003). Monkeys reject unequal pay. *Nature*, 425(6955), 297–299.

Chang, S. W., Winecoff, A. A., & Platt, M. L. (2011). Vicarious reinforcement in rhesus macaques (Macaca mulatta). *Frontiers in Neuroscience*, 5(27), 1–10.

Chen, M. K., Lakshminarayanan, V., & Santos, L. R. (2006). How basic are behavioral biases? Evidence from capuchin monkey trading behavior. *Journal of Political Economy*, 114(3), 517–537.

Dar-Nimrod, I., & Heine, S. J. (2011). Genetic essentialism: On the deceptive determinism of DNA. *Psychological Bulletin*, 137(5), 800.

Delgado, M. R., Schotter, A., Ozbay, E. Y., & Phelps, E. A. (2008). Understanding overbidding: Using the neural circuitry of reward to design economic auctions. *Science*, 321(5897), 1849–1852.

Devine, P. G., & Elliot, A. J. (1995). Are racial stereotypes really fading? The Princeton trilogy revisited. *Personality and Social Psychology Bulletin*, 21(11), 1139–1150.

Fiske, S. T. (2010). Interpersonal stratification: Status, power, and subordination. In S. T. Fiske, D. T. Gilbert, & G. Lindzey (Eds.), *Handbook of social psychology* (5th ed., Vol. 2, pp. 941–982). Hoboken, NJ: Wiley.

Gilbert, G. M. (1951). Stereotype persistence and change among college students. *Journal of Personality and Social Psychology*, 46, 245–254.

Guinote, A. (2007). Power and goal pursuit. *Personality and Social Psychology Bulletin*, 33(8), 1076–1087.

Harris, L. T. (2014). Why economic, health, legal, and immigration policy should consider dehumanization. *Policy Insights from the Brain and Behavioral Science*, 1, 1–7. doi:10.1177/2372732214548593.

Jayaratne, T. E., Ybarra, O., Sheldon, J. P., Brown, T. N., Feldbaum, M., Pfeffer, C. A., et al. (2006). White Americans' genetic lay theories of race differences and sexual orientation: Their relationship with prejudice toward blacks, and gay men and lesbians. *Group Processes & Intergroup Relations*, 9(1), 77–94.

Karlins, M., Coffman, T. L., & Walters, G. (1969). On the fading of social stereotypes: Studies in three generations of college students. *Journal of Personality and Social Psychology*, 13(1), 1–16. doi:10.1037/h0027994.

Katz, D., & Braly, K. (1933). Racial stereotypes in one hundred college students. *Journal of Abnormal and Social Psychology*, 28, 280–290.

Maslach, C. (1982). *Burnout: The cost of caring*. Englewood Cliffs, NJ: Prentice-Hall.

Pinker, S. (2011). *The better angels of our nature: The decline of violence in history and its causes*. London: Penguin.

Singer, P. (1975). *Animal liberation: A new ethics for our treatment of animals*. New York: Random House.

Vaes, J., & Muratore, M. (2012). Defensive dehumanization in the medical practice: A cross-sectional study from a health care worker's perspective. *British Journal of Social Psychology*, 52, 180–190. doi:10.1111/bjso.12008.

Williams, M. J., & Eberhardt, J. L. (2008). Biological conceptions of race and the motivation to cross racial boundaries. *Journal of Personality and Social Psychology*, 94(6), 1033.

Appendix: A Brief Guide to the Brain

Given the interdisciplinary nature of the conversation, it behooves me to spare a quick word for technical details about the nervous system, including the brain. The nervous system controls the entire body. The central nervous system extends from neurons that transmit information from the retina and other primary sensory organs to the brain, to neurons that transmit information from the brain to the spinal column. The peripheral nervous system comprises the spinal column and the neurons that extend from it to the muscles throughout the body (for detailed discussion, see Gazzaniga, 2004). The brain evolved to support survival and achieves this through movement. The example of the sea squirt (urochordata) best captures this sentiment. As an adult, it resembles a tube, filtering water through the body and gaining nutrients. However, when young, the creature looks like a tadpole; it has eyes, a brain, and fins. It searches for a nutrient-rich environment, finds it, and then digests its eyes and brain, becoming essentially a stomach. This underlies the role of the brain in promoting survival through movement. However, the human brain does so much more than govern our behavior; it is unique perhaps because it creates a sense of self and other conscious experiences, and can flexibly engage social cognition. Human brains can imagine what is not there, create art and culture, and solve complex problems. Nonetheless, it would serve one well to remember this basic lesson from the sea squirt.

The brain comprises a subcortex, a phylogenetically older structure shared with many other species, and a neocortex, the phylogenetically newer part of the brain. Although many other species have a neocortex, the human's is the most complex, with brain regions not found in other species. This difference separates us from other animals. I discuss structures in both the neo- and subcortices. However, the brain does not operate in a modular function; it is a collection of networks, transmitting electrochemical signals that generate thought, action, emotion, and decisions. Therefore, brain regions are described in the context of networks. Although this is not the level of detail required to properly characterize every neuronal function, it is sufficient for describing more abstract social phenomena.

Finally, there is a general principle that describes brain function that is worth keeping in mind. The brain is not a hard drive—it does not store exact copies of files in your head, and it does not accurately perceive anything. What the brain does is recreate past events, not recall them, and fill in gaps in perception with expectations. This is all in service of constructing a coherent reality. Numerous cognitive researchers have been playing tricks on the brain for years, demonstrating both functions of the organ (see for one such discussion, Orlandi, 2012). For instance, the McGurk effect (McGurk & MacDonald, 1976) demonstrates that visual input can influence audition. By pairing the auditory component of one syllable with the visual component (e.g., seeing an actor's lips move) or another sound, a third sound is perceived. This feature of the brain dovetails well with the cognitive consistency idea in social psychology—human beings strive for both internal and external consistency of thoughts and behaviors (Simon & Holyoak, 2002). Theories such as cognitive dissonance (Festinger & Carlsmith, 1959) and stereotype threat (Steele, 1997) also describe the brain's/mind's attempt to construct a coherent whole by changing either attitudes or behavior. This feature of the brain is relevant to flexible social cognition because it serves the same function: shaping expectations for a coherent world consistent with one's beliefs.

References

Festinger, L., & Carlsmith, J. M. (1959). Cognitive consequences of forced compliance. *Journal of Abnormal and Social Psychology*, 58, 203–210.

Gazzaniga, M. S. (2004). *The cognitive neurosciences*. Cambridge, MA: MIT Press.

McGurk, H., & MacDonald, J. (1976). Hearing lips and seeing voices. *Nature*, 264, 746–748.

Orlandi, N. (2012). Visual switching: The illusion of instantaneity and visual search. *Review of Philosophy and Psychology*, 3(4), 469–480. doi:10.1007/s13164-012-0098-z.

Simon, D., & Holyoak, K. J. (2002). Structural dynamics of cognition: From consistency theories to constraint satisfaction. *Personality and Social Psychology Review*, 6(4), 283–294.

Steele, C. M. (1997). A threat in the air: How stereotypes shape intellectual identity and performance. *American Psychologist*, 52(6), 613.

Index